Praise for *Math Workshop Essentials: Developing Number Sense Through Routines, Focus Lessons, and Learning Stations*...

The lessons, routines, and strategies in this outstanding resource support exactly the kind of teaching we need in our classrooms to build students' number sense and overall confidence in math. I will be putting it in the hands of every teacher in my district!

—Andy Johnsen, superintendent
Lakeside Union School District, Lakeside, California

When we know better, we do better. And the ideas in Bresser and Holtzman's resource help us all know better by providing us practical activities to do with students.

—Jennifer Lempp, author
Math Workshop: Five Steps to Implementing Guided Math,
Learning Stations, Reflection, and More

This resource has positively influenced me as a math tutor and a supervisor of math practicum students at the university level. The authors have provided powerful mathematical content while thoughtfully guiding educators to engage students in meaningful mathematical conversations. It's outstanding!

—Vicki Bachman, math tutor and adjunct professor
University of Iowa, Iowa City, Iowa

Rusty and Caren's thoughtful and inspiring lesson designs provide highly accessible and motivating support to my math teachers. The activities offer rich examples for engaging a full range of learners in thinking and talking about mathematics.

—Kathleen Gallagher, principal
Baker Elementary School, San Diego, California

Clear and accessible, this resource gently encourages us to be reflective in our math teaching. What I like most about the resource is that it centers on teacher and student thinking and discussion in a way that brings the classroom to life. Whether you are just starting out in teaching or a seasoned professional, you don't want to miss this wonderful resource!

—Edwin Mayorga, assistant professor of educational studies
Swarthmore College, Swarthmore, Pennsylvania

This resource illustrates the importance of developing number sense in children and offers engaging, easy-to-follow lessons that deepen student understanding and empower young mathematicians.

—Mary Ann Warrington, math specialist and coach
Stuart Hall for Boys, San Francisco, California

Math Workshop Essentials

About This Series

This series is intended to offer content for the three math workshop structures featured in *Math Workshop* (Lempp 2017, 183).

MATH WORKSHOP: TASK AND SHARE		MATH WORKSHOP: FOCUS LESSON, GUIDED MATH, AND LEARNING STATIONS			MATH WORKSHOP: GUIDED MATH AND LEARNING STATIONS		
5–10 minutes	NUMBER SENSE ROUTINE	5–10 minutes	NUMBER SENSE ROUTINE		5–10 minutes	NUMBER SENSE ROUTINE	
30 minutes	MATH TASK	15 minutes	FOCUS LESSON		45 minutes	GUIDED MATH	LEARNING STATIONS
		30 minutes	GUIDED MATH	LEARNING STATIONS			
20–25 minutes	TASK SHARE WITH STUDENT REFLECTION	5–10 minutes	STUDENT REFLECTION		5–10 minutes	STUDENT REFLECTION	

MATH WORKSHOP ESSENTIALS

Developing Number Sense Through Routines, Focus Lessons, and Learning Stations

GRADES 3–5

Rusty Bresser
Caren Holtzman

Foreword by Jennifer Lempp

Math Solutions
Boston, Massachusetts, USA

Math Solutions
www.mathsolutions.com

Library of Congress Cataloging-in-Publication Data

Names: Bresser, Rusty, author. | Holtzman, Caren, author.
Title: Math workshop essentials : developing number sense through routines, focus lessons, and learning stations / Rusty Bresser, Caren Holtzman.
Description: Boston, Massachusetts, USA : Math Solutions, a division of Houghton Mifflin Harcourt, [2018] | Includes bibliographical references.
Identifiers: LCCN 2018024759 | ISBN 9781935099598 (pbk.)
Subjects: LCSH: Mathematics—Study and teaching (Elementary)
Classification: LCC QA135.6 .B7339 2018 | DDC 510.71—dc23
LC record available at https://lccn.loc.gov/2018024759

ISBN-13: 978-1-935099-59-8
ISBN-10: 1-935099-59-0

Math Solutions is a division of Houghton Mifflin Harcourt.

Oh No! 99! is adapted from a UNO Company game, now out of print.

Tell Me All You Can is adapted from *Smart Arithmetic, Grades 4–6*, by Rhea Irvine and Kathryn Walker (Creative Publications).

Get to 1,000, Get to Zero, and *Hit the Target* are adapted from Calculators in Mathematics Education (NCTM Yearbook) (Reston, VA: National Council of Teachers of Mathematics, 1992).

Decimal Maze Game Board is reprinted with permission of the National Council of Teachers of Mathematics.

Executive Editor: Jamie Ann Cross
Production Manager: Denise A. Botelho
Cover design: Wanda Espana, Wee Design Group
Cover photo: Friday's Films (from *Math Workshop* by Jennifer Lempp, 2017)
Interior design and composition: Publishers' Design and Production Services, Inc.

Printed in the United States of America.

1 2 3 4 5 6 7 8 9 10   0928   26 25 24 23 22 21 20 19 18
4510005258

A Message from Math Solutions

We at Math Solutions believe that teaching math well calls for increasing our understanding of the math we teach, seeking deeper insights into how students learn mathematics, and refining our lessons to best promote students' learning.

Math Solutions shares classroom-tested lessons and teaching expertise from our faculty of professional learning consultants as well as from other respected math educators. Our publications are part of the nationwide effort we've made since 1984 that now includes

- more than five hundred face-to-face professional learning programs each year for teachers and administrators in districts across the country;

- professional learning books that span all math topics taught in kindergarten through high school;

- videos for teachers and for parents that show math lessons taught in actual classrooms;

- on-site visits to schools to help refine teaching strategies and assess student learning; and

- free online support, including grade-level lessons, book reviews, inservice information, and district feedback, all in our Math Solutions Online Newsletter.

For information about all of the products and services we have available, please visit our website at *www.mathsolutions.com.* You can also contact us to discuss math professional learning needs by calling (877) 234-7323 or by sending an email to *info@mathsolutions.com.*

We're always eager for your feedback and interested in learning about your particular needs. We look forward to hearing from you.

*We dedicate this book to our mentor, Marilyn Burns,
who is a constant inspiration and whose work reminds
us that learning mathematics should be an enjoyable,
sense-making endeavor.*

Contents

Foreword by Jennifer Lempp   ix

Acknowledgments   xi

How to Use This Resource   xiii

Connections to the Common Core State
Standards   xxiii

Activities Categorized by Number Sense Feature   xxxv

Section I   **Number Sense Routines**

Engaging, accessible, purposeful routines
to begin math workshop and more   1

Routine 1: Tell Me All You Can   3

Routine 2: Number Talks   10

Routine 3: Number Lines   19

Routine 4: Guess My Number   30

Routine 5: Computational Estimation   43

Section II   **Focus Lessons**

Whole-class lessons to use in math
workshop and more   55

Focus Lesson 1: Trail Mix   57

Focus Lesson 2: Numbers and Me   68

Focus Lesson 3: How Many Beans?   77

Focus Lesson 4: Fraction Ballparks   88

Section III   **Learning Stations (Games)**

Games for learning stations in math
workshop and more   97

Game 1: One Time Only   99

Game 2: Oh No! 99!   110

Game 3: Hit the Target   122

Game 4: Get to Zero   133

Game 5: Get to 1,000 (Addition)   143

Game 6: Get to 1,000 (Multiplication)   153

Game 7: Decimal Maze   163

References   173

Reproducibles   175

Foreword

Math workshop provides differentiated instruction for students and offers them engaging learning experiences that promote mathematical thinking, discourse, and a positive disposition toward mathematics. Mistakes are a part of the process in math workshop, and teachers and students alike view these mistakes as important in the learning process. Math workshop decreases anxiety by creating a safe and engaging environment where students enjoy working with numbers, persevere through the toughest of problems, and respect one another. When students no longer feel anxiety, their engagement increases and they see themselves as capable problem solvers.

This resource, as part of the Math Workshop Essentials series, gives educators valuable guidance for beginning their math workshop journeys. It offers routines, lessons, and games that increase engagement and are high quality, meaningful, and thoughtful. The content is timeless and easily fits into the three structures of math workshop that I encourage educators to use. In addition, Bresser and Holtzman's resource doesn't just "feed" readers step-by-step directions; the authors have taken the time to share insightful experiences they have had teaching each and every activity. They instill the confidence that you too can jump into math workshop! The number sense routines, focus lessons, and learning stations can be used time and time again. The descriptions, directions, and examples are friendly and accessible for teachers of all levels of experience with mathematics. By making these activities part of your math instruction, you will see the power that quality mathematics teaching has on student engagement.

> By making these activities part of your math instruction, you will see the power that quality mathematics teaching has on student engagement.

Every teacher who reads this resource has a way to get started with the math workshop model today. When we know better, we do better. And the ideas in Bresser and Holtzman's resource help us all know better by providing us practical activities to do with students. We all got into education with one priority in mind—the student. These authors enthusiastically share that same philosophy and put students at the heart of their work.

—Jennifer Lempp, author of *Math Workshop: Five Steps to Implementing Guided Math, Learning Stations, Reflection, and More*

Acknowledgments

We thank the following people for making this book possible: Jamie Cross, Mary Mitchell, and Denise Botelho for their guidance and support with the book's development and production. We'd also like to thank:

Andrea Barraugh

Dina Calvin

Shea Carillo

Julie Contestable

Ann Dominick

Carolyn Felux

Marji Freeman

Kathleen Gallagher

Robin Gordon

Sally Haggerty

Judy Kozak

Pam Long

Lynn Lorimer

Lyndsey Lovelace

Deb Obregon

Annette Raphel

Patti Reynolds

Kathy Seckington

Christina Stamford

Serena Thakur

Maryann Wickett

Pam Wilson

And special thanks to Cesar Bohorquez and Elaina Taylor.

How to Use This Resource

While there is no simple checklist of skills that number sense encompasses, we recognize the characteristics of number sense and their effects. We see our students use their number sense when they are confronted with numerical situations. We hear them describe their number sense when they explain how they solved a problem.

> "A person who possesses number sense might be said to have a well-integrated mental map of a portion of the world of numbers and operations and is able to move flexibly and intuitively throughout the territory."—Paul R. Trafton (1991)

This resource shares some of the ways we have attempted to help students "move throughout the territory." It offers practical and worthwhile ideas for helping students think about numbers through number sense routines, focus lessons, and learning stations. It will help you understand number sense and find innovative ways to promote it in your classroom. And it will show you how teaching number sense can facilitate procedural fluency and math reasoning, two important goals for standards-based instruction.

What Is Number Sense?

Read the following scenarios. For each one, ask yourself: *How are the students and/or adults exhibiting number sense?*

Scenario 1: Amanda's Understanding of 27 × 4

In a third-grade classroom, Ms. Mendoza asks her students to mentally figure the answer to 27 × 4. She gives them a few moments to solve the problem, and then directs the students to turn to a neighbor and share their answer and how they got it. Amanda explains to her partner, David, "First, I thought of four quarters and did 25 times 4 which is a dollar, or 100. Then I had two left from the 27, so I did 2 times 4. A hundred plus 8 is 108." After David explains his strategy, Amanda says excitedly, "Oh! I have another way! You could do 20 times 4, then 7 times 4, and then add 80 and 28!

Scenario 2: Felipe's Understanding of 31 × 11

Mr. Lao's fourth graders are telling him all that they know about the answer to 31 × 11. Felipe says, "I know the answer's going to be greater than 300, because 30 times 10 is 300, so it's got to be more than that."

Scenario 3: Karin's Understanding in Comparing $\frac{1}{2}$ and $\frac{3}{8}$

Karin, a fifth grader, works with her partner, Kyle, comparing the fractions that are written on little cards. Karin places a card on the table that reads $\frac{1}{2}$, and Kyle puts down a card with $\frac{3}{8}$. Karin tells Kyle, "One-half is bigger, because $\frac{4}{8}$ is the same as $\frac{1}{2}$, and $\frac{3}{8}$ is less than $\frac{4}{8}$."

Scenario 4: Marie's Understanding of 10 percent of $13.00

A group of adults are just finishing their dinner at a restaurant when the waiter delivers the check. The bill totals $138.56. As John reaches for his calculator, his wife Marie says, "What are you doing? Ten percent is about $13.00, so for a 20 percent tip, just leave double that."

Number sense is a broad idea that covers a range of numerical thinking. Although the concept can be difficult to pinpoint, we recognize number sense when we see our students use it. The children and adults in these scenarios tap in to their number sense when confronted with problems or when thinking about numbers and operations. In Scenario 1, Amanda demonstrates her flexibility with numbers and operations and an understanding of place value and number decomposition as she mentally solves 27 × 4 in two different ways. She also reveals her awareness of the different uses for numbers in the world by making a connection to money. In Scenario 2, Felipe is able to use computational estimation to get a sense of the reasonableness and magnitude of an answer. In Scenario 3, Karin makes use of what she already knows about equivalency in order to compare two fractions. And in Scenario 4, we see that even as adults, we often rely on our number sense when solving problems in everyday life as evidenced by Marie's way of reasoning about how much of a tip to leave.

Why Number Sense? (The Four Features)

We have identified four features that we feel are integral to the development of and need for number sense. All the activities in the book help students

1. compute mentally,
2. compute fluently,
3. navigate our number system with understanding, and
4. effectively use estimation.

Icons indicate which number sense features are showcased in each activity. Additionally, a Number Sense Features table on page xxxv shows activities listed by number sense focus areas. Following are descriptions of each of these important number sense features.

Mental Computation MC

Mental computation is a key aspect of number sense. Students with number sense can manipulate numbers in their head. They have strategies they use to think about numbers and operations. They do not need to rely solely on paper and pencil, their fingers, or a calculator. Mental computation forces students to rely on what they know about numbers and operations by liberating them from the standard paper-and-pencil approach to computation and allowing them to be more inventive.

Computing mentally shifts the emphasis from following procedures to making sense of numbers and operations. It allows students to develop and use their own algorithms, which may be more efficient and make use of important ideas such as place value. When students develop their mental computation skills they deepen their understanding of the relationships between and among numbers. They think flexibly about numbers, are able to break numbers apart and put them together in a variety of ways. They have an understanding of place value and can use this knowledge when operating on numbers. They are also familiar with the properties of single-digit numbers and can use this information to calculate efficiently using larger numbers.

Procedural Fluency (PF)

Students with number sense are able to compute fluently. This means that they have a strong grasp of single-digit number facts, which are the foundation of computation. They have a repertoire of strategies or procedures from which to choose when solving problems, and these strategies are ones that make sense to them.

Students with procedural fluency understand the effects operations have on numbers. They see connections between the different operations and have a firm grasp of which operation or series of operations to employ in a given situation. They are able to articulate why they choose a particular operation and how it will help them solve the problem.

Students can use their procedural fluency to transfer their understandings to new situations. They know which methods are appropriate in various problem-solving situations and over time become more efficient and accurate with the methods they choose and use.

While procedural fluency is important, students also need latitude in their thinking about numbers if they are to develop number sense. Repetitive exercises with isolated numbers do not suffice. Students need to be able to make sense of what they are doing. When students share their own computational strategies, everyone in the class benefits from hearing a variety of approaches. The act of listening and making sense of someone else's approach to numbers and computation forces the listeners to expand their horizons. As students hear about and understand more ways to think about numbers, they augment their own mathematical abilities. They become more flexible thinkers who have more than one way to deal with a novel problem when it arises, and they remain curious and enthusiastic learners.

> "Effective teaching of mathematics builds fluency with procedures on a foundation of conceptual understanding so that students, over time, become skillful in using procedures flexibly as they solve contextual and mathematical problems."
> —NCTM (2011)

Navigating the Number System (N)

In addition to becoming comfortable with number facts and numerical operations, students need experiences that give them a broader understanding of our base ten number system, a system that consists of very orderly patterns and consistent characteristics. This place-value system applies to both whole and decimal numbers. Upper-grade students need experiences that unify their place-value concepts. For example, we expect students to know, without using paper and pencil or a calculator, what happens when 10 is added to a number or when a number is multiplied by 10. Understanding the importance of tens and tenths in our number system gives students access to many computational strategies.

Similarly, students need to recognize the patterns inherent in odd and even numbers, factors, and multiples. These relationships serve them in making reasonable predictions and finding efficient computation strategies. These understandings also help students make sense of fractions, their relative quantities, and equivalent forms. If children are aware of the structure and patterns behind our number system, they are able to understand relationships among numbers and can predict and evaluate reasonable solutions to math problems. This deeper understanding and the ability to predict are essential components of number sense.

Teachers play a critical role in guiding students' learning about the number system. The key is to give students many opportunities to negotiate the system, letting them chart their own course through familiar territory and challenging them to discover and investigate new paths along the way.

Estimation E

Students with number sense have effective ways to estimate. They can approximate calculations and use familiar benchmarks to gauge unknown amounts. We estimate every day. Do I have enough spaghetti to feed the six people who are coming for dinner? After I pay the rent and my bills, can I afford a weekend getaway? About how much punch should I buy for my daughter's birthday party? We answer these questions through a combination of mental computation and estimation. We don't usually pull out a calculator in order to figure the tip in a restaurant. We don't use paper and pencil to decide whether we have enough time to stop at the grocery store before meeting a friend for lunch. Estimating also helps us spot an answer that doesn't make sense. When the cashier rings up a dozen donuts and the total comes to over fifty dollars, we both know something is wrong.

It's extremely important that students have many opportunities to estimate, because the skill develops over time and with experience. The types of practice, discussion, and thinking associated with estimating help build number sense. Estimating helps students think of numbers as quantities, and opportunities to think about numbers in context are key. Context makes numbers real. It is also important for students to be able to decide when accuracy is essential and when an estimate will be good enough (or even better).

As students gain skill and experience with estimating in different contexts, they bring more and more number sense to tasks. They learn to use benchmarks: *I know how many beans are in the small scoop so I can use that information to estimate how many beans are in the larger scoop.* They begin to get a feel for quantities that are reasonable: *There are 30 students in our class, and it*

The National Research Council (2001) recommends "the mathematics curriculum provides opportunities for students to develop and use techniques for mental arithmetic and estimation as a means of promoting a deeper number sense "

looks like there are enough chairs in the auditorium for 5 classes. They develop a sense of relative magnitude: *If I have 150 cookies, will 8 serving plates be enough to hold them all?*

How Is This Resource Organized? (For Math Workshop and More)

Though the ideas in this resource can be used for various types of instructional models, they are primarily organized with the use of math workshop in mind.

In this resource, we focus on three components that are within the structures of math workshop:

1. Number Sense Routines
2. Focus Lesson
3. Learning Stations (Games)

Let's look more closely at each of these components and the format.

For more on this model of instruction, see *Math Workshop* by Jennifer Lempp (Math Solutions, 2017).

The Three Sections

Number Sense Routines

A number sense routine is defined by Lempp as "an engaging, accessible, purposeful routine to begin your math class that promotes a community of positive mathematics discussion and thinking" (2017, 104). Number sense routines are sometimes referred to as "minilessons" or "warm-ups" and provide students with repeated practice over the course of the school year. These low-prep ideas are an alternative to more traditional practice with worksheets and are typically teacher-led activities that engage students in thinking, computing, communicating, and problem solving. In some cases, students can take charge and lead the routine once they become familiar with the format and expectations. Having students lead an activity communicates a message of shared ownership of the math workshop time.

Once well established, routines take only five to ten minutes of class time. However, we strongly recommend spending a longer period when first introducing the routines to students. Taking the time to model the routine with the students, elicit strategic thinking and solution strategies, and clarify any procedures and language demands will save time in the long run. Investing time in introducing the routines means that students will be comfortable and competent with the structure of the routine and will be ready to dive in quickly in subsequent use.

Focus Lessons

A focus lesson is defined by Lempp as "a well-planned, whole-group lesson focused on the day's learning target and accessible to all levels of learners" (2017, 121). These activities lend themselves to deeper exploration and connections and are structured in a "launch, explore, summary" format. They include a description of how the teacher introduced the lesson to the class, followed by a description of student work time. Making use of student reflection in summarizing these rich whole-class math tasks is the key to bringing thinking and number sense concepts to the surface. Detailed descriptions of how the teacher summarized the lesson with a deliberate use of questions offers the reader a variety of ways to bring students' number sense to the forefront of the activity.

Learning Stations (Games)

Defined by Lempp as "activities in which students engage in meaningful mathematics and are provided with purposeful choices" (2017, 121), the games for learning stations serve as hybrids of routines and focus lessons. Number sense games help students develop and practice number sense skills in an engaging and thought-provoking context. In most instances, the games are introduced using the whole-class lesson format. As with routines, the initial investment in a whole-class introduction to the game pays off when students have the understanding and experience to play independently. The opportunities to learn a game, play it with a partner, analyze structure and strategy, and listen to classmates' thinking turn the games into rich math tasks worthy of a math class period.

One of the great things about the games is that they can live on in the classroom long after the introductory lesson is over. Versions of the games can be incorporated into math workshop time and can also be used as routines and productive options for early finishers.

The Format

In order to help you with planning and teaching the ideas in this resource, each activity is presented in an easy-to-follow format.

Overview

To help you decide if the lesson is appropriate for your students, the "Overview" is a nutshell description of the mathematical goal of the lesson and what the students will do.

Learning Targets

Learning targets are identified for each grade level, making it easier to set specific grade-level, standards-based content objectives.

Materials

This section includes special materials needed (aside from the board, paper, and pencils), along with quantities.

Time

The number sense routines, intended to begin math workshop, typically are five to ten minutes in length. The focus lessons can take up to forty-five minutes and in some cases can be extended to multiple days. The games in the learning stations section typically take fifteen to thirty minutes when students are playing independently. The introduction to any new math routine, game, or structure merits more time as students learn new content, structure, procedures, and vocabulary.

Teaching Directions

The directions are presented in a concise step-by-step format (and also interwoven with the vignettes in the "From the Classroom" section).

Modifications/Extensions

For the focus lessons and some of the learning stations, follow-up activities are suggested, as well as ideas for modifying or extending the lessons to make them appropriate for different grade levels and abilities.

From the Classroom

Vignettes that describe what actually occurred when the lesson was taught to one or more classes are included in each lesson. While the vignettes mirror the steps summarized in the teaching directions, they elaborate with details that are valuable for preparing and teaching the lesson. In each vignette, important mathematical practices are highlighted to help teachers see examples of these practices in action, and to help them understand how number sense plays a role in developing expertise with the practices. Samples of student work are included.

Reflecting on the Activity

This section offers insights about number sense, teaching decisions, grade-level appropriateness, and suggestions about assessing student understanding.

Sidenotes

In addition to the lesson's components, special tips occasionally appear in the margin. Key features of the lesson are highlighted and they denote areas of particular emphasis. There are three tip categories:

1. **Mathematical Practices**
   In this sidenote, sometimes we point out connections to the Mathematical Practices in the Common Core State Standards (see page xxiii). These practice standards illuminate how students engage in the mathematics. While the content standards help determine learning targets for each grade level, the practice standards guide teachers in thinking about how to structure lessons and questions to maximize students' opportunities to do and make sense of mathematics.

2. **Formative Assessment**
   We identify formative assessment opportunities throughout the text. Often teachers can learn a great deal about students' number sense by asking a well-timed question or choosing a specific prompt or work product. Formative assessment teaches us about our students' thinking and understanding and helps guide our instructional decisions.

3. **Teaching Tip**
   The final sidenote is the Teaching Tip. This catchall category draws attention to teacher moves and techniques that we have found particularly effective in our math teaching experiences. Often the subtler layers of a lesson go unnoticed because managing the lesson and the students' responses in real time requires a great deal of attention. The Teaching Tips slow down the process and give a window into some of the underlying microdecisions that happen during the lesson.

How Do I Get Started?

There is no prescribed way to use the book or the activities: this is not a program, curriculum, or sequential unit. Rather, it is a spectrum of ways to foster number sense in the intermediate grades, no matter what state standards or textbook you are using to guide your teaching of mathematics. If you are just beginning to focus on number sense in your classroom, we hope you will find here practical ideas and insights into the richness and power of number sense. If you are already focusing on number sense, we hope you will find some activities and perspectives to add to your repertoire.

The math workshop approach lends itself well to these activities. We suggest you use the activities in ways that best meet the needs of your students, and we encourage you to make adaptations as you go about bringing number sense to the forefront of your mathematics instruction. For those teaching the Common Core State Standards, see pages xxiii–xxxiii for connections. Like number sense, teaching itself develops over time with support and experience. We hope you find this resource a useful tool for your professional practice and for your students' number sense.

Connections to the Common Core State Standards

Where Do We See Number Sense in the Standards?

Students with number sense exhibit the same characteristics as those described by the Common Core State Standards for mathematically proficient students. In the Common Core State Standards for Mathematical Practice, proficient students are identified as being able to:

- **Explain** to themselves the meaning of problems and look for entry points to its solution.
- **Check** their answers to problems using a different method, and they continually ask themselves, "Does this make sense?"
- **Apply** the mathematics they know to solve problems arising in everyday life, society, and the workplace.
- **Justify** their conclusions, communicate them to others, and respond to the arguments of others.
- **Interpret** their mathematical results in the context of the situation and reflect on whether the results make sense.
- **Detect** possible errors by strategically using estimation and other mathematical knowledge.
- **Calculate** accurately and efficiently, express numerical answers with a degree of precision appropriate for the problem context.
- **Evaluate** the reasonableness of their results.

These are precisely the varieties of expertise that we seek to develop in students through the activities and strategies in this book. Examples of these key practices in action are highlighted throughout the activity vignettes. Additionally, the lessons in the book help students gain important experience with key skills and ideas from the Common Core Content Standards that are highlighted in each activity, such as:

- using place value understanding and properties of operations to perform multidigit arithmetic;
- developing an understanding of fractions as numbers;
- understanding the properties of multiplication and the relationship between multiplication and division;
- understanding decimal notation and comparing decimal numbers;
- understanding the place value system, and
- performing operations with fractions and decimals.

The activities in this book help students meet these Common Core State Standards, with a focus that fosters number sense development. Teachers can help students make this shift by asking questions and employing strategies that encourage them to estimate, visualize quantities, make connections, and compute mentally.

No longer is it sufficient for students to solve problems using procedures they learned by rote. The Common Core State Standards call for a balance between understanding concepts and procedural fluency. "Mathematical understanding and procedural skill are equally important, and both are assessable using mathematical tasks of sufficient richness" (NGA Center & CCSSO 2010). The lessons in this book that are designed to help students develop their number sense are examples of these rich mathematical tasks.

What About the Standards for Mathematical Practice?

While the content standards specify what math content is expected at each grade level, the practice standards define how students are expected to engage in mathematics across the grade levels. The following Standards for Mathematical Practice (NGA Center & CCSSO 2010) are highlighted throughout the activities in this book in order to provide teachers with concrete examples.

1. **MP1. Make sense of problems and persevere in solving them.**

   Mathematically proficient students

   - *explain* to themselves the meaning of a problem and looking for entry points to its solution;
   - *analyze* givens, constraints, relationships, and goals;
   - *make conjectures* about the form and meaning of the solution and plan a solution pathway rather than simply jumping into a solution attempt;
   - *monitor* and evaluate their progress and change course if necessary;
   - *explain* correspondences between equations, verbal descriptions, tables, and graphs or draw diagrams of important features and relationships;
   - *check* their answers or problems using a different method; and
   - *continually ask themselves*, "Does this make sense?"

2. **MP2. Reason abstractly and quantitatively.**

   Mathematically proficient students

   - *make sense of* quantities and their relationships in problem situations;
   - *abstract* a given situation and represent it symbolically;
   - *attend to* the meaning of quantities, not just how to compute them; and
   - *know and flexibly use* different properties of operations and objects.

3. **MP3. Construct viable arguments and critique the reasoning of others.**

   Mathematically proficient students

   - *understand and use* stated assumptions, definitions, and previously established results in constructing arguments;
   - *make conjectures* and build a logical progression of statements to explore the truth of their conjectures;

- *justify* their conclusions, communicate them to others, and respond to the arguments of others;
- *construct arguments* using concrete referents such as objects, drawings, diagrams, and actions; and
- *listen to or read* the arguments of others, decide whether they make sense, and ask useful questions to clarify or improve the arguments.

4. **MP4. Model with mathematics.**

Mathematically proficient students

- *apply* the mathematics they know to solve problems arising in everyday life;
- *identify* important quantities in a practical situation and map their relationships using such tools as diagrams, two-way tables, graphs, flowcharts, and formulas, and analyze those relationships mathematically to draw conclusions; and
- *interpret* mathematical results in the context of a situation and reflect on whether the results make sense, possibly improving the model if it has not served its purpose.

5. **MP5. Use appropriate tools strategically.**

Mathematically proficient students

- *consider* the available tools when solving a problem (e.g., paper and pencil, concrete models, a ruler, a calculator); and
- *use* technological tools to explore and deepen their understanding of concepts.

6. **MP6. Attend to precision.**

Mathematically proficient students

- *communicate* precisely to others;
- *use* clear definitions in discussion with others and in their own reasoning;
- *state* the meaning of the symbols they choose, including using the equal sign consistently and appropriately;
- *calculate* accurately and efficiently, and express numerical answers with a degree of precision appropriate for the problem context; and
- *give* carefully formulated explanations to each other.

7. **MP7. Look for and make use of structure.**

Mathematically proficient students

- *look closely for* patterns or structure; and
- *can step back for an overview* and shift perspective when problem solving.

8. **MP8. Look for and express regularity in repeated reasoning.**

Mathematically proficient students

- *notice* if calculations are repeated;
- *look for* general methods or shortcuts;
- *maintain oversight* of the process when problem solving, while attending to details; and
- *evaluate* the reasonableness of their intermediate results when solving a problem.

The complete standards can be found at: www.corestandards.org/Math.

The following tables show how the activities connect to the Common Core content and practice standards.

Third Grade

CONNECTIONS TO THE CONTENT STANDARDS

	Tell Me All You Can	Number Talks	Number Lines	Guess My Number	Computational Estimation	Trail Mix	Numbers and Me	How Many Beans?	Fraction Ballparks	One Time Only	Oh No! 99!	Hit the Target	Get to Zero	Get to 1,000 (Addition)	Get to 1,000 (Multiplication)	Decimal Maze
Operations and Algebraic Thinking																
Represent and solve problems involving multiplication and division.	X	X		X	X			X		X			X		X	
Understand properties of multiplication and the relationship between multiplication and division.	X	X						X		X						
Multiply and divide within 100.	X	X		X	X			X		X		X	X		X	
Solve problems involving the four operations and identify and explain patterns in arithmetic.	X	X			X			X		X	X	X	X	X	X	
Numbers and Operations in Base Ten																
Use place-value understanding and properties of operations to perform multidigit arithmetic.	X	X		X	X			X			X	X	X	X	X	
Numbers and Operations—Fractions																
Develop an understanding of fractions as numbers.	X		X				X	X	X							

Third Grade

CONNECTIONS TO THE STANDARDS FOR MATHEMATICAL PRACTICE

	Tell Me All You Can	Number Talks	Number Lines	Guess My Number	Computational Estimation	Trail Mix	Numbers and Me	How Many Beans?	Fraction Ballparks	One Time Only	Oh No! 99!	Hit the Target	Get to Zero	Get to 1,000 (Addition)	Get to 1,000 (Multiplication)	Decimal Maze
1. Make sense of problems and persevere in solving them.	X	X		X	X			X						X	X	
2. Reason abstractly and quantitatively.		X	X				X	X		X	X					
3. Construct viable arguments and critique the reasoning of others.		X	X	X			X		X		X	X		X	X	
4. Model with mathematics.	X		X													
5. Use appropriate tools strategically.					X							X	X			
6. Attend to precision.	X	X		X	X			X	X	X	X					
7. Look for and make use of structure.		X		X						X				X	X	X
8. Look for and express regularity in repeated reasoning.		X		X						X			X	X	X	

Fourth Grade

CONNECTIONS TO THE CONTENT STANDARDS

	Tell Me All You Can	Number Talks	Number Lines	Guess My Number	Computational Estimation	Trail Mix	Numbers and Me	How Many Beans?	Fraction Ballparks	One Time Only	Oh No! 99!	Hit the Target	Get to Zero	Get to 1,000 (Addition)	Get to 1,000 (Multiplication)	Decimal Maze
Operations and Algebraic Thinking																
Use the four operations with whole numbers to solve problems.	X	X		X	X			X		X	X	X	X	X	X	
Gain familiarity with factors and multiples.		X		X				X		X		X	X		X	
Generate and analyze patterns										X		X	X			
Numbers and Operations in Base Ten																
Generalize place-value understanding for multidigit whole numbers.	X	X			X			X				X		X	X	
Use place-value understanding and properties of operations to perform multidigit arithmetic.	X	X			X			X				X		X	X	
Numbers and Operations—Fractions																
Extend understanding of fraction equivalence and ordering.	X		X	X	X				X							
Build fractions from unit fractions.	X		X			X			X							
Understand decimal notation for fractions and compare fraction decimals.		X	X				X									

Fourth Grade

CONNECTIONS TO THE STANDARDS FOR MATHEMATICAL PRACTICE

	Tell Me All You Can	Number Talks	Number Lines	Guess My Number	Computational Estimation	Trail Mix	Numbers and Me	How Many Beans?	Fraction Ballparks	One Time Only	Oh No! 99!	Hit the Target	Get to Zero	Get to 1,000 (Addition)	Get to 1,000 (Multiplication)	Decimal Maze
1. Make sense of problems and persevere in solving them.	X	X		X	X	X		X						X	X	
2. Reason abstractly and quantitatively.		X	X	X		X	X	X		X	X					
3. Construct viable arguments and critique the reasoning of others.		X		X			X		X	x	X	X		X	X	
4. Model with mathematics.	X		X		X											
5. Use appropriate tools strategically.					X							X	X			
6. Attend to precision.	X	X		X	X			X	X	X	X					
7. Look for and make use of structure.		X		X					X				X	X	X	
8. Look for and express regularity in repeated reasoning.		X		X						X			X	X	X	X

Fifth Grade

CONNECTIONS TO THE CONTENT STANDARDS

	Tell Me All You Can	Number Talks	Number Lines	Guess My Number	Computational Estimation	Trail Mix	Numbers and Me	How Many Beans?	Fraction Ballparks	One Time Only	Oh No! 99!	Hit the Target	Get to Zero	Get to 1,000 (Addition)	Get to 1,000 (Multiplication)	Decimal Maze
Operations and Algebraic Thinking																
Write and interpret numerical expressions.																X
Analyze patterns and relationships.					X					X		X	X			X
Numbers and Operations in Base Ten																
Understand the place-value system.	X	X		X	X						X	X	X	X	X	X
Perform operations with multidigit whole numbers with decimals to hundredths.	X	X			X											X
Numbers and Operations—Fractions																
Use equivalent fractions as a strategy to add and subtract fractions.	X	X	X		X	X			X							
Apply and extend previous understandings of multiplication and division.	X	X	X		X	X			X							

Fifth Grade

CONNECTIONS TO THE STANDARDS FOR MATHEMATICAL PRACTICE

	Tell Me All You Can	Number Talks	Number Lines	Guess My Number	Computational Estimation	Trail Mix	Numbers and Me	How Many Beans?	Fraction Ballparks	One Time Only	Oh No! 99!	Hit the Target	Get to Zero	Get to 1,000 (Addition)	Get to 1,000 (Multiplication)	Decimal Maze
1. Make sense of problems and persevere in solving them.	X	X		X	X	X		X						X	X	X
2. Reason abstractly and quantitatively.		X	X	X		X	X	X		X	X					X
3. Construct viable arguments and critique the reasoning of others.		X		X			X		X		X	X		X	X	
4. Model with mathematics.	X		X	X		X										
5. Use appropriate tools strategically.					X							X	X			X
6. Attend to precision.	X	X			X			X	X	X	X					
7. Look for and make use of structure.		X		X						X				X	X	X
8. Look for and express regularity in repeated reasoning.		X		X						X			X	X	X	

Activities Categorized by Number Sense Feature

This table categorizes the activities (routines, focus lessons, and learning stations) in this resource by their number sense feature. The larger, bolder X's indicate the activity's most prominent number sense feature. For more on the four number sense features, see page xv of the "How to Use This Resource" section.

	Number Sense Feature 1: Mental Computation (MC)	Number Sense Feature 2: Procedural Fluency (PF)	Number Sense Feature 3: Navigating the Number System (N)	Number Sense Feature 4: Estimation (E)
Routines				
Tell Me All You Can		**X**		X
Number Talks		**X**		
Number Lines		X	**X**	
Guess My Number			**X**	
Computational Estimation	X			**X**
Focus Lessons				
Trail Mix		**X**		X
Numbers and Me			**X**	X
How Many Beans?	X			**X**
Fraction Ballparks			X	**X**

(continued)

	Number Sense Feature 1: Mental Computation (MC)	Number Sense Feature 2: Procedural Fluency (PF)	Number Sense Feature 3: Navigating the Number System (N)	Number Sense Feature 4: Estimation (E)
Learning Station (Games)				
One Time Only		X	X	
Oh No! 99!	X	X		
Hit the Target		X		X
Get to Zero	X	X	X	
Get to 1,000 (Addition)	X	X		
Get to 1,000 (Multiplication)	X	X	X	
Decimal Maze		X	X	

Number Sense Routines

Section Overview

A number sense routine is defined by Jennifer Lempp as "an engaging, accessible, purposeful routine to begin your math class that promotes a community of positive mathematics discussion and thinking" (2017, 104). Number sense routines are sometimes referred to as "minilessons" or "warm-ups" and provide students with repeated practice over the course of the school year. These low-prep ideas are an alternative to more traditional practice with worksheets and are typically teacher-led activities that engage students in thinking, computing, communicating, and problem solving. In some cases, students can take charge and lead the routine once they become familiar with the format and expectations. Having students lead an activity communicates a message of shared ownership of the math workshop time.

Once well established, routines take only five to ten minutes of class time. However, we strongly recommend spending a longer period when first introducing the routines to students. Taking the time to model the routine with the students, elicit strategic thinking and solution strategies, and clarify any procedures and language demands will save time in the long run. Investing time in introducing the routines means that students will be comfortable and competent with the structure of the routine and will be ready to dive in quickly in subsequent use.

ROUTINES

R-1	Tell Me All You Can	3
R-2	Number Talks	10
R-3	Number Lines	19
R-4	Guess My Number	30
R-5	Computation Estimation	43

Tell Me All You Can

Overview

Computing mentally, making estimates, and exploring relationships among numbers all help students develop number sense. In *Tell Me All You Can*, students tell all they can about the answers to a series of arithmetic problems. While they may know the exact answer to a problem, the activity requires them to think about ways to describe the answer using concepts such as close to, between, greater than, and less than.

Learning Targets

GRADE 3

I can determine how reasonable my answers are using mental computation, estimation, and rounding.

GRADE 4

I can determine how reasonable my answers are using mental computation, estimation, and rounding.

I can use number sense and fractions that I know to estimate the reasonableness of answers to fraction problems.

GRADE 5

I can determine how reasonable my answers are using mental computation, estimation, and rounding.

I can use number sense and fractions that I know to estimate the reasonableness of answers to fraction problems.

NUMBER SENSE FEATURES

(see page xv)

3

Materials

None

 **Time**

Ten to fifteen minutes

Teaching Directions at a Glance

1. Write an arithmetic problem on the board using a horizontal format (for example: 30×11).

2. After giving students time to think about the problem, ask them what they can say about the answer without actually revealing the answer (regarding 30×11, for example: "I think the answer will be more than 300 because 30 times 10 is 300").

3. After eliciting as many responses as possible, repeat Steps 1 and 2 with another arithmetic problem.

Teaching Directions with Classroom Insights

From a Fifth-Grade Classroom

Before posing arithmetic problems for students to solve, I posted the following sentence frames on the board and had the class practice reading them aloud. The frames are intended to help students, especially those learning English as a second language, frame responses when asked to explain their thinking—an important mathematical practice in the Common Core State Standards.

▸ The answer is going to be around/about _____ because _____.

▸ The answer is going to be close to _____ because _____.

> The answer is going to be between _____ and _____ because _____.

> The answer is going to be greater than _____ because _____.

> The answer is going to be less than _____ because _____.

1. Write an arithmetic problem on the board using a horizontal format.

After practicing the sentence frames with the class, I wrote the following problem on the board horizontally, since the vertical format often triggers the algorithm of starting from the right and "carrying." I wanted students to look at the numbers as a whole and to think about the largest parts of the numbers first.

2. After giving students time to think about the problem, ask them what they can say about the answer without actually revealing the answer.

I then asked the students, "Without figuring the exact answer, what are some things you know about the answer?"

$$12 \times 7$$

"I think the answer is going to be less than 120, because 12 times 10 is 120," Damarie said.

"I think the answer is going to be greater than 60 because I know that 12 times 5 is 60," added Jimmy.

"I think it's going to be more than 12 times 6, because that's six 12s and 12 times 7 is seven 12s," Miriam explained.

These three students were able to calculate accurately and efficiently, a key mathematical practice.

3. After eliciting as many responses as possible, repeat Steps 1 and 2 with another arithmetic problem.

I wrote another problem on the board:

$$30 \times 11$$

"I think it will be more than 300, 'cause 30 times 10 is 300," Kathy reasoned. Students who use their number sense will often look at a problem

Teaching Tip

Encourage students to calculate accurately and efficiently; this is an important part of number sense.

holistically before confronting details. Instead of focusing on individual digits, Kathy first thought about 30, then she multiplied by 10, which is close to but less than 11, yielding a pretty good approximation of the answer.

"It's going to be less than 400," Chrissy added.

"Why do you think that?" I probed, trying to elicit an explanation.

"Because I know that 30 times 10 is 300 and there's only one more 30, which is less than 100," she reasoned.

The next problem I wrote on the board challenged students to break numbers apart even further.

$$75 \times 12$$

This problem drew "oohs" and "ahhs" from students and seemed to pose more of a challenge for them. When no one raised a hand, I asked a question to prompt some thinking: "Will the answer be more or less than 100?"

"It's going to be a lot bigger than 100!" Amanda exclaimed. "Seventy-five times 2 is more than 100, and the problem is 75 times 12!"

When no one else raised their hand, I had the students talk with partners about their thoughts, and then brought them back together for a discussion. Partner talks give students time to think and chat; it's like rehearsing before sharing with the class.

"I think it's going to be a lot bigger than 150, because 75 times 2 is 150, and you have to go 10 times bigger than that," Miguel reasoned.

"It's easy," Dave said. "You just do 75 times 10, and you get 750, then you do two more 75s and add that to 750. It'll be between 800 and 900."

Next, I switched things up by posing a fraction problem:

$$\frac{1}{2} + \frac{1}{3}$$

Teaching Tip

Encourage students to make reasonable approximations; this is an important part of having number sense.

"It's going to be bigger than $\frac{3}{4}$, because $\frac{1}{2}$ is $\frac{2}{4}$, and a third is bigger than a fourth," Courtney explained. Mathematically proficient students who can apply what they know ("$\frac{1}{2}$ is $\frac{2}{4}$, and $\frac{1}{3}$ is bigger than $\frac{1}{4}$") are comfortable making assumptions and approximations (the answer is "going to be bigger than $\frac{3}{4}$"). Making reasonable approximations is an important part of having number sense.

"Do you think the answer is going to be greater or less than a whole?" I asked the class.

"I think the answer's going to be less than 1, because $\frac{1}{2}$ plus $\frac{1}{2}$ is a whole, and $\frac{1}{2}$ is bigger than $\frac{1}{3}$," Dave argued. He came up to the board and drew a picture of a cookie divided first into thirds, and then into halves to compare $\frac{1}{3}$ and $\frac{1}{2}$. Mathematically proficient students like Dave can use their number sense to identify quantities and map their relationships ($\frac{1}{2}$ compared to $\frac{1}{3}$) using tools like diagrams and drawings.

Teaching Tip

Use tools like diagrams and drawings to help students identify quantities and map their relationships.

From a Third-Grade Classroom

1. Write an arithmetic problem on the board using a horizontal format.

To start off the lesson, I posed the following problem on the board:

$$\$10.00 - \$1.99$$

2. After giving students time to think about the problem, ask them what they can say about the answer without actually revealing the answer.

"The answer's gonna be less than $10.00, 'cause you take money away from ten dollars," Tiffany said.

"It's going to be around $8.00, because $1.99 is only a penny less than $2.00, and 10 minus 2 is 8," Andy explained.

These responses impressed me; the students were using their knowledge about operations and working with friendly numbers, both indicators of number sense. Next, I wrote a three addend expression:

$$45 + 45 + 45$$

Using one of the sentence frames for support, Jesycha said, "It's gonna be between 100 and 150!"

"What made you think of numbers between 100 and 150?" I asked, fishing for an explanation.

"Because 45 plus 45 is 90, and there's another 45, so the answer's gonna be somewhere in between 100 and 150," she explained.

"I'm thinking of it like 45 times 3, because 45 times 3 is the same as 45 plus 45 plus 45," Manuel said. Number sense enables students to see the relationships between operations. This often involves making conjectures about the form and meaning of problems, an important mathematical practice.

Teaching Tip

Making conjectures can help students make sense of the form and meaning of problems.

3. After eliciting as many responses as possible, repeat Steps 1 and 2 with another arithmetic problem.

I followed up with another subtraction problem that included *49*, a number that could easily be rounded to a friendlier number.

$$49 - 25$$

I called on Elba, who showcased her number sense by renaming the problem to make the estimation easier. "The answer's going to be around 25," she said. "I changed the 49 into a 50, and I know that 50 minus 25 is like 25 cents."

Andy went next. "It's gonna be under 50, because 49 is close to 50 and you take away another 25," he explained.

"The answer's going to be the same as 12 times 2," Dan added, already having solved the problem mentally and finding another way to express it.

Reflecting on the Lesson

What is the purpose of this activity?

Tell Me All You Can works for almost any grade level to give students practice with arithmetic while building their number sense. In an activity like this, students have opportunities to think about reasonableness, place value, and number meaning. They also benefit from the chance to acquire new computation and estimation strategies.

Why is an activity like this better than procedural drills?

Children benefit from frequent practice solving arithmetic problems, but the practice they get in this activity asks them to think about the numbers involved and what happens to these numbers when they are added, subtracted, multiplied, and divided. Rather than learning a certain procedure and practicing it over and over, students are encouraged to think about the reasonableness of their estimates and explain their reasoning. This kind of practice builds a student's number sense.

How can I make sure this activity will be effective?

It's important to pose a variety of problems, some easy, some more challenging. This allows more students to have access to the thinking that's required. Students also need time to think about the problems and talk with one another in order to clarify their ideas and get more than one perspective. Students always need to explain their reasoning, so that others can benefit from their thinking. Finally, as the teacher, you need to think about the problems beforehand and develop possible questions that will stimulate students' thinking.

How can I use this activity for assessment purposes?

Tell Me All You Can gave Sally Haggerty an idea of the range of understanding her fifth graders had about fractions. This was valuable to her since she hadn't yet begun teaching them about fractions. Knowing what experience and understanding students bring to a topic in math is important, because when we learn something new, we build on what we already know. Sometimes what we know makes mathematical sense and sometimes it doesn't. An activity like *Tell Me All You Can* can alert you to students' misunderstandings or lack of experience, whether it's fifth graders dealing with fractions or third graders making sense of subtraction.

Many of the students you called on seem to already have good number sense. How do you help those who don't?

There are many supportive strategies a teacher can use to help students develop their number sense. During *Tell Me All You Can*, partner talks gave all students a chance to think, explain their thinking to a partner, and listen to someone else's idea. Questions such as, "Do you think the answer will be greater than or less than a whole?" or "Do you think the answer will be more or less than 100?" provide parameters that help students focus their thinking. Giving mathematical hints can also help students. For example, when talking about 45 + 45 + 45, a teacher might ask, "What number is close to 45 that might help you think about the answer?" And finally, providing a safe environment where mistakes are seen as learning sites and risk taking is encouraged goes a long way in supporting students.

Formative Assessment

Tell Me All You Can can alert you to students' strengths, misunderstandings, or lack of experience.

ROUTINE

Number Talks

Overview

Teachers have many options in determining when and how to use number talks. Since they are usually brief (ten- to fifteen-minute conversations), they can be peppered into any lesson and used throughout the year. Number talks work as a warm-up activity at the beginning of a lesson. They can arise from a problem or context within a lesson. They might be incorporated into a summary discussion. In any situation, number talks develop students' insight, understanding, and communication skills.

While seemingly simple, number talks actually require a good deal of practice and skill to facilitate. This routine provides some strategies and tools to support the use of number talks as an integral part of number sense development.

There are examples of number talks scattered throughout the sections of this resource. These number talks illustrate how teachers can integrate the approach in the context of a math lesson. This routine slows down the process and deconstructs a few number talks in order to deeply analyze effective implementation.

**NUMBER SENSE
FEATURES**

(see page xv)

Learning Targets

GRADE 3

I can solve problems using multiplication and division.

I can use what I know about place value and operations to solve problems with bigger numbers.

GRADE 4

I can use the four operations to solve problems.

I can use place value to operate on large numbers.

GRADE 5

I can solve math equations with large numbers.

I can solve math equations with decimals.

I can use the four operations with fractions.

Materials

None

 **Time**

Ten to fifteen minutes

Teaching Directions at a Glance

1. Write a computation problem on the board.
2. Have students solve the problem mentally.
3. Have partners share their solution strategy.
4. Solicit different solutions strategies from the students.
5. Record the different strategies on the board, connecting student descriptions with proper mathematical notation.

Teaching Directions with Classroom Insights

From a Third-Grade Classroom

Number strings consist of a sequence of related problems that build on one another. Third-grade students focus heavily on building foundations in multiplication and division. Using number strings in number talks can help students see connections between problems and learn how to use what they know about smaller problems to help them solve larger problems.

For example:

$$4 \times 5 =$$

$$4 \times 50 =$$

$$4 \times 5 \times 10 =$$

$$20 \times 10 =$$

The teacher prompts students to use what they know about the preceding problem to help solve the subsequent problem. This number string simultaneously keeps students thinking about the operation of multiplication and the base ten number system.

While multiplication is a major third-grade focus, the students also need practice with all four operations, and they need to continue exploring the place-value system and how it works with larger numbers. Offering number talks with multidigit addition and subtraction keeps third graders' numbers and operations skills sharp.

The following are two examples of third-grade number talks.

Mathematical Practice

Connecting problems and using relationships help students reason abstractly and quantitatively (MP2).

450 + 126 =

Jennifer's Way

450 + 126 =

126 = 100 + 20 + 6
450 + 100 = 550
550 + 20 = 570
570 + 6 = 576

Oscar's Way

450 + 126 =

400 + 100 = 500
50 + 20 = 70
500 + 70 = 570
570 + 6 = 576

Ali's Way

450 + 126 =

450 + 100 = 550
550 + 26 = 576

Katie's Way

450 + 126 =

6 + 0 = 6
50 + 20 = 70
400 + 100 = 500
500 + 70 + 6 = 576

312 − 97 =

Dan's Way

312 − 97 =

97 $\xrightarrow{+3}$ = 100
312 − 100 = 212
212 + 3 = 215

Leonor's Way

312 − 97 =

97 + 3 = 100
100 + 200 = 300
300 + 12 = 312
3 + 200 + 12 = 215

Joey's Way

$$312 - 97 =$$

$$97 + \boxed{100} = 197$$
$$197 + 100 = 297$$
$$297 + 3 = 300$$
$$300 + 12 = 312$$
$$100 + 100 + 3 + 12 = 215$$

The students benefit from thinking about, hearing, and seeing different ways multidigit numbers can be decomposed and recombined to accommodate various operations.

From a Fourth-Grade Classroom

In fourth grade, students spend a lot of time operating on multidigit numbers. They use what they've learned about the place-value system to make sense of computation with larger numbers. Number talks are crucial in helping students keep the focus on sense making and helpful strategies. In addition to solid place-value foundations, using friendly numbers and decomposing and recombining numbers are vital strategies for understanding computation with larger numbers.

The distributive property is an offshoot of the decomposing and recombining that students begin in the earlier grades. Using the distributive property helps simplify multidigit problems into manageable chunks based on place value and knowledge of single-digit facts. For example, when fourth graders mentally solved 998×3, several of the students saw that 998 was close to 1,000 so they added 2 to 998 and multiplied by 3. Then they needed to subtract out the extra twos that were included in the initial computation. Essentially, they did the following:

$$(3 \times 1,000) - (3 \times 2)$$

This idea can be simplified to $3(1,000 - 2)$. So, the students were informally using the concept of the distributive property to solve the problem.

Regular use of number talks in the classroom allows students to develop their flexibility and fluency with numbers and operations. While initially some students have difficulty moving past the standard algorithm that was drilled into their heads, they eventually can be freed to find more efficient and sensible ways to operate on numbers.

The following is another example:

$$75 \times 12 =$$

Yvette's Way

75 × 12 =

75 × 10 = 750
75 × 2 = 150
750 + 150 = 900

A.J.'s Way

75 × 12 =

70 × 12 = 940
5 × 12 = 60
840 + 60 = 900

Kylie's Way

75 × 12 =

70 × 10 = 700
70 × 2 = 140 } 840
5 × 10 = 50
5 × 2 = 10 } 60 } 900

When students construct their own understandings and strategies, they are able to use the properties of operations with ease and comfort, rather than merely following a procedure or rule.

From a Fifth-Grade Classroom

Fifth graders work intensively on fractions and decimal numbers. As they explore this terrain, they need to maintain their understanding of quantity and symbols. Often students jump to paper-and-pencil computation and symbol manipulation. Having them "just think" about a question without paper and pencil gives rise to fascinating discussions. A rich number talk can develop from simple comparison problems such as:

Which is more?

$$\frac{1}{6} \text{ or } \frac{2}{8}$$

Fifth graders also do multidigit multiplication and division problems. In these circumstances, number talks can help them to build on their place value understanding and look for ways to decompose and recombine factors, dividends, and divisors.

$$1{,}250 \div 50 =$$

Marco's Way

1,250 ÷ 50 =

Money!
How many 50 cents are in $12.50?
24 in $12
1 in .50
24 + 1 = 25

Prianka's Way

1,250 ÷ 50 =

50 × 2 = 100 (30)
50 ×⃝20 = 1,000
250 ÷ 50 =⃝5
20 + 5 = 25

Diana's Way

1,250 ÷ 50 =

1,000 ÷ 50 = 20
250 ÷ 50 = 5
20 + 5

Michael's Way

1,250 ÷ 50 =

125 ÷ 5 = (100 ÷ 5) + (25 ÷ 5)
125 ÷ 5 = 20 + 5 = 25

Money serves as a nice context for number talks involving decimal numbers. Asking students to estimate gauges how well they conceptualize quantities less than zero and how they deal with "unfriendly" numbers.

Julia went shopping. Here's a list of her expenses. Did she spend more than $12 or less than $12? How do you know?

$6.37 $1.25 $2.65

When I presented this problem to students I was encouraged to see that most of them started with the larger numbers to get a sense of the total. Then they used landmark decimal numbers (0.25, 0.50, 0.75) to estimate the value of the remaining cents.

Number talks like this can be used as a dipstick to check the class's general comfort and facility with certain types of numbers and operations. The discussions also give insight into an individual student's thinking and quantitative reasoning approaches. Routinely seeking out chances to have brief number talks with the class provides teachers with invaluable information while building students' number and operation sense.

Formative Assessment

Number talks give immediate information about how students think about numbers and computation.

Reflecting on the Lessons

What are the main purposes of number talks?

Number talks focus on computation, comparison, or estimation problems. They engage both the social and the visual modes of thinking and learning. Teachers use number talks to showcase quantitative reasoning and make different approaches both public and visible. The power of number talks resides in

- Opportunities to solve the problem mentally;
- Opportunities to communicate thinking;
- Opportunities to solve problems in more than one way;
- Opportunities to listen to and consider other ways to solve the problem; and
- Opportunities to record student thinking symbolically.

What are some tips for facilitating successful number talks?

Students (and teachers) get better at number talks with practice and time. There's no avoiding a learning curve. However, there are some practical tips that can move number talks into richer, more engaging territory. Following is advice to assist the teacher just starting to institute number talks in the classroom. These suggestions are nice reminders for veterans of number talks as well.

1. *Write the problems horizontally.* This simple act opens the discussion up to a world of possibilities. If the problems are written vertically with one number lined up above the other, most students will automatically jump to the standard algorithm. By writing the numbers in a horizontal format, students are free to think about and look at the numbers in a variety of ways.

2. *Start easy.* Number talks require a high cognitive load (computation, reflection, communication, listening), so in order to help students succeed, start with easy problems. Initially the focus is on helping students understand the format and norms for a number talk. Students also need to learn how to explain their thinking and listen to others' ideas. That's a lot to do. It's best not to start with a problem that's at the edge of their independent computation abilities. Let the computation itself be easy so the students can practice organizing their thinking, talking, listening, and responding to questions. Once they get those basics, gradually ramp up the computation challenges.

3. *Write the problem each time a student describes a new strategy.* It might seem redundant to repeat the problem, but it's very helpful for students. Start fresh each time a student shares his or her thoughts. That way, everyone can follow the entire solution path from start to finish.

4. *Use sequencing words to make sure equations are recorded accurately.* It's tempting to record students' thought processes in exactly the way they describe them. But sometimes their descriptive language doesn't match proper mathematical notation. For example, in solving $147 + 254$, a student said, "I added 147 and 200 and I got 347, plus 50 more is 397, plus 3 more is 400, plus 1 more is 401."

   Taking this description verbatim, I might have written:

   $$147 + 200 = 347 + 50 = 397 + 3 = 400 + 1 = 401$$

   But looking closely at the chain, it's obvious that those expressions are not all equal. The equal sign is not a connector between ideas; it's an indicator that the quantities on both sides are the same. Clearly, $147 + 200$ is not the same as $347 + 50$, so instead I wrote the student's thinking this way:

   $$147 + 200 = 347$$

   $$347 + 50 = 397$$

   $$397 + 3 = 400$$

   $$400 + 1 = 401$$

   Using sequencing prompts helps me keep the steps of the process clear and also helps slow down the student's description so I can follow and record accurately. I ask questions such as, "What did you do first? Where did those numbers come from?" Then "What did you do?"

5. *First do the problem yourself. Anticipate possible responses.* Sometimes it's hard to make sense of a student's explanation. The better prepared you are to think about the problem in different ways, the more likely it is that you'll recognize your students' thinking. Doing number talks with friends and family is a good way to practice facilitating and prepare for different responses. (This suggestion presumes your friends and family are loving and patient.)

   But also know that your students will inevitably blow you away by thinking of brilliant ways you hadn't considered. That's one of the joys of number talks. Conversely, there are occasions where it's just not possible to follow a student's line of thinking. When this occurs, I'll ask other students if they can help with the explanation. Sometimes I just have to tell the student that I'm a little confused and I'll get back to them later. (And then I need to make sure that I do follow-up.)

6. *Keep students engaged.* Pacing and active engagement are crucial to successful number talks. Students will check out mentally if they are held hostage to a long conversation volley that only involves the teacher and one other student. There are many ways to encourage active listening and participation. These strategies include having students

   - explain their thinking to a partner;
   - try someone else's strategy;
   - explain their partner's strategy;
   - use thumbs up or down to show if they agree or disagree with a response; and
   - ask questions.

7. *Step back and let the students lead.* Once students are familiar with the norms and protocols of number talks, they can take the lead. Students can come to the front of the classroom and talk about their strategies while recording the corresponding equations on the board. They can lead discussions and ask other students questions. They can also look for connections among different strategies and solution paths.

Related Resources

For more on number talks, see the following resources:

> *Number Talks: Whole Number Computation* by Sherry Parrish (Math Solutions, 2010, 2014)

> *Number Talks: Fractions, Decimals, and Percentages* by Sherry Parrish and Ann Dominick (Math Solutions, 2016)

Number Lines

Overview

Number lines are an invaluable tool for helping students develop their number sense, because they offer a visual model of quantitative reasoning and relationships. They serve addition and subtraction well. They're a great tool for number talks and helping make students' thinking visible. An open number line is a tool used for capturing the steps taken to solve a computation problem. A number line can also be used to help students conceptualize the constant difference strategy.

There are several different types of number lines with different uses. In addition to aiding with computation, number lines can help students consider the relative magnitude of numbers (including fractions and decimals) and their relationships to one another. Many upper-elementary students struggle with comparing fractions. The number line gives them a linear model to add to their repertoire.

NUMBER SENSE FEATURES

(see page xv)

Learning Targets

GRADE 3

I can add and subtract numbers within 1,000.

I can use place value to round to the nearest ten or hundred.

GRADE 4

I can determine how reasonable my answers are by using estimation, mental math, and rounding.

I can compare large numbers by using what I know about place value.

GRADE 5

I can compare two decimals and use <, >, = symbols.

I can estimate the reasonableness of answers to fraction problems.

Materials

None

 ## Time

Ten to twenty minutes

Teaching Directions at a Glance

INTRODUCING CONSTANT DIFFERENCE

1. Facilitate a number talk. For more on number talks, see Routine 2 in this section.
2. Use a number line to model students' thinking
3. Show the constant difference strategy as appropriate.

THINKING ABOUT BASE TEN AND PLACE VALUE

1. Draw a number line segment with the two end points labeled.
2. Write some numbers on the board.
3. Have students decide where the numbers would fit on the number line.
4. Ask them to justify their reasoning.

Teaching Directions with Classroom Insights

From a Third-Grade Classroom: Introducing Constant Difference

1. Facilitate a number talk.

I did a number talk with a third-grade class. I chose a relatively simple problem because I wanted to introduce a new strategy—constant difference.

I had students mentally solve *74 – 26*. The students had a variety of ways to solve the problem but I zeroed in on Javier, who shared that he rounded 26 up to 30 by adding 4. Then he added 4 to 74 and got 78. Then he subtracted 30 from 78.

2. Use a number line to model students' thinking.

I used a number line to show what Javier had done. (See Figure R3–1.)

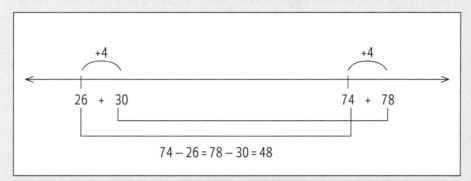

Figure R3–1. Using the constant difference strategy

3. Show the constant difference strategy as appropriate.

I demonstrated that it's OK to slide the entire problem up the number line as long as the distance between the two numbers remains constant (hence, *constant difference*). The constant difference strategy gives students a very visual way to think about and compare quantities.

The constant difference strategy is handy in a number of ways. First, it eliminates the need to regroup. Regrouping often becomes the focus of traditional subtraction instruction. In many ways the manipulation of single digits reduces students' use of place value and their understanding of the operation of subtraction. This algorithmic emphasis on "borrowing and carrying" actually detracts from students' number sense.

Another nice feature of the constant difference strategy is that it presents a model of subtraction as comparison, not take-away. It's important for students to have exposure to different types of subtraction problems. In addition, the constant difference approach provides a practical reason for rounding to the nearest ten.

Sometimes the tricky part of the constant difference strategy is choosing the number to round. An easy rule of thumb is to always round the subtrahend. (In the expression 74 − 26, 74 is the minuend and 26 is the subtrahend.) If you can round the subtrahend to a ten, it will be easier to compare to any number. Interestingly, rounding up or rounding down works equally well with constant difference. As long as the minuend and subtrahend move the same amount and in the same direction on the number line, the difference will be the same as well.

Mathematical Practice

An open number line is a handy tool that students can strategically use to solve problems (MP5).

I often introduce the constant difference strategy in the context of a number talk. I look for a student like Javier, who describes a mental computation strategy that mirrors constant difference. The number line gives a very visual model of how it works. Because I know it's probably a new strategy for many students, I give them some time to grapple with it and let them mess around with a few more problems.

In the third-grade class, I had students work with a partner to try to use the constant difference strategy on a number line. I scaffolded the problems so they became progressively more challenging with larger numbers. I also tried to offer a range of problems. In some cases, rounding up would be more efficient, while rounding down would work well in others. I let the students make the decisions, but it was interesting to see how they tackled the tasks. Here are some problems I used:

$$56 - 38 =$$
$$70 - 42 =$$
$$125 - 87 =$$

The constant difference strategy can take some time to make sense of and get comfortable with, but it's well worth the investment. It helps students build their number sense by thinking of numbers as whole quantities and fluidly moving around the number system.

From a Fourth-Grade Classroom: Thinking About Base Ten and Place Value

A focus of the fourth-grade year is operating on multidigit numbers. Number lines can help students maintain their number sense in the midst of procedural computation. Number lines can help remind students of the relative magnitude of numbers and their relationships to one another. In using number lines to compare and order numbers, students also practice justifying their thinking and explaining why they put a number in a chosen spot on the line.

I began by writing some sentence frames on the board:

I think _____ goes between _____ and _____ because _____.

It will be closer to _____ because _____.

It goes to the right/left of _____ because it is greater than/less than _____.

Then, because I was introducing a new tool and new vocabulary, I chose a relatively easy problem to start. I wanted to make sure the students could be successful and I also didn't want them to struggle with the math so much that they didn't have room in their brains for the sentence frame or thinking about the number line.

1. Draw a number line segment with the two end points labeled.

I drew a 1–1,000 number line on the board:

←————————————————————————————————————→

1   1,000

2. Write some numbers on the board.

I wrote these numbers on the board:

500, 200, 750, 600, 300

3. Have students decide where the numbers would fit on the number line.

"Where does 500 go?" I asked the class. "Remember to use the sentence frame to explain your answer."

4. Ask them to justify their reasoning.

"I think 500 goes right in the middle because it's half of 1,000," Rene offered.

"OK," I replied, "Any other ideas? Even if you agree, did you have another way of thinking about it?"

"It's like a ruler," Jonny explained. "and 500 is ½ of 1,000."

"Yes," I agreed. "Sometimes I think about number lines as rulers or racetracks. Building on Jonny's idea, if this number line were a 1,000-meter race track, how far would I be when I was half of the way across?"

"Five hundred meters," several students responded.

"Good," I assented. "Not that I run a lot of 1,000-meter races, but you never know."

I put *500* in the middle of the number line.

←————————————————————————————————————→

1   500   1,000

Assured that the students had the language and the strategies to move on, I told them to talk with a partner about where they would place 200,

750, 600, and 300 on the number line. I encouraged them to use the sentence frames and make sure they agreed with each other.

After students worked in pairs for a few minutes, we got back together for a whole-group discussion. We co-constructed the number line and put the numbers where they belonged. Along the way, I pushed students to justify their reasoning, ask each other questions, and make sure there was agreement before plotting any of the points. Again, I reminded them of the available sentence frames to help in constructing their arguments.

Then I drew a 1–10,000 number line, which offered more of a challenge.

"OK, this time you're going to think about a new number line. This time think about where 1,000 goes. It's a bit tricky, so really think and listen to your partner's ideas, too."

1   10,000

Teaching Tip

Start with a simpler problem or provide an accessible benchmark; this opens the door for more students to access the problem.

"Where does 1,000 go on this number line?" I asked.

I let the students talk to their partners and could tell almost immediately there was some confusion. Understanding place value and the relative magnitude of tens, hundreds, and thousands might seem like something fourth graders would have in hand. However, many students have lots of experiences with procedures and manipulating digits, not necessarily thinking about larger quantities and how they compare.

I decided to give them more support by asking where 5,000 would go. They were quite comfortable putting 5,000 in the middle of the line, and I encouraged them to use that knowledge to think about where 1,000 fit.

"Can anyone use a sentence frame to help us figure out where 1,000 goes?" I prompted.

"It goes to the left of 5,000 because it's less than 5,000," Angela responded.

"OK," I pointed, "so we know it's somewhere on this side of the number line. But where?"

"It's closer to 0 because there's 1,000, 2,000, 3,000, and 4,000 before you get to 5,000," Markus added.

"It's like fractions," Julianna noticed.

"How so?" I asked.

"Because there are ten 1,000s in 10,000. So 1,000 is only one-tenth of 10,000. If you divided the line into ten pieces, 1,000 would be the first piece. It will be much closer to 0 than 5,000."

"Yeah," Tom agreed, "5,000 is a half and 1,000 is only one-tenth."

I was pleased with this brief exchange. Students were thinking about place-value relationships, relative magnitude, and a linear model of fractions. That's the kind of richness that helps develop number sense, and that's why number lines are powerful tools.

From a Fifth-Grade Classroom: Fractions on a Number Line

1. Draw a number line segment with the two end points labeled.

In Deb Obregon's class, I started by drawing a number line:

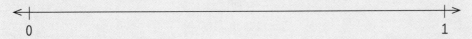

0   1

(On a side note, I once drew the same number line in a third-grade classroom. One of the girls was astonished. She said, "I didn't know there were any numbers between zero and one!")

2. Write some numbers on the board.

I then wrote the following numbers on the board:

$$\frac{1}{2}, \; \frac{1}{3}, \; \frac{2}{3}$$

3. Have students decide where the numbers would fit on the number line.

I began by asking Deb's fifth graders where $\frac{1}{2}$ went (no-brainer). Then we talked about where $\frac{1}{3}$ would go. After a bit of discussion, the students agreed.

"OK," I continued. "Talk to a partner about where $\frac{2}{3}$ would go."

4. Ask them to justify their reasoning.

I presented the same sentence frames I had used with the fourth-grade class. The discussions were productive, and it didn't take long for the students to have solid arguments for where to put 2/3.

I then gave the students a task to work on with a partner.

"Now you're going to compare two fractions. You're going to figure out which is more and then decide where each one fits on the number line. Talk it over with your partner and also think of some ways you can prove it. You can also use paper and pencil if you have ideas about how it might help." Having students prove their answer gave them opportunities to engage in constructing viable arguments and critiquing the reasoning of others.

I differentiated the activity by giving students some choices. Each pair had to do the first comparison. Then I allowed them to choose an "interesting challenge" to try.

Which is more? Where do these fractions go on the number line?

Everyone:

$$\frac{1}{2} \text{ or } \frac{1}{8}$$

Choose a Challenge:

$$\frac{5}{5} \text{ or } \frac{8}{10}$$

$$\frac{1}{6} \text{ or } \frac{8}{12}$$

$$\frac{4}{5} \text{ or } \frac{3}{6}$$

$$\frac{1}{3} \text{ or } \frac{4}{5}$$

$$\frac{4}{4} \text{ or } \frac{3}{6}$$

Deb had her students write in their math journals, giving her an assessment opportunity. Students used a variety of approaches to compare the fractions. Many, like Alexandra, used an area model to compare the fractions. (See Figure R3–2.)

Some students used a part-whole method and compared the number and size of pieces in each fraction. Antonio explained his thinking. (See Figure R3–3.)

Once students decided which fraction was larger, they worked together to decide where each fraction fit on the number line. This simple activity turned out to be very rich. It gave the students opportunities to compare fractions, put fractions on a number line, and develop the language and logic mathematics.

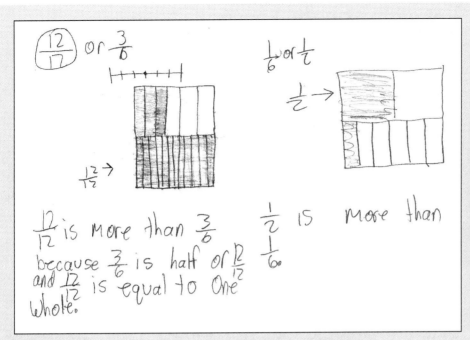

Figure R3–2.   Alexa used the area model to compare fractions.

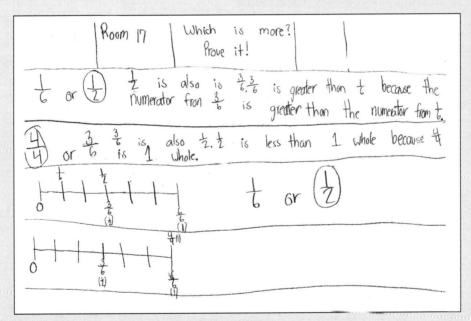

Figure R3–3.   Antonio constructed a viable argument in writing and illustrated his thinking with number lines.

Reflecting on the Lesson

How can we help students become more comfortable with number lines?

In general, the main issue with number lines is unfamiliarity. A number line is a tool and students become comfortable with it by using it. The more we can pepper number lines into our math instruction, the more students will come to understand their structures and uses. Asking questions such as *Where does ___ go? How do you know? Is there a friendly number close by?* helps students understand that the number line is a way to model the number system.

Another tip is to not always start at 0. In fact, when you get down to it, theoretically number lines don't "start" anywhere. The number line is a geometric (linear) model of the number system; technically it goes on infinitely in both directions. We're just choosing a particular segment of it to use in our classrooms. So why not mix it up sometimes? For example, present students with a 100–600 number line segment and ask them where 400 fits. These tasks can lead to rich conversations and also develop the habit of looking at both ends of the number line segment to orient in the proper part of the number system.

The "Constant Difference" example on page 21 uses an open number line. The open number line reinforces flexible use of number lines and requires students to pay attention to "where they are." Many teachers have not used number lines regularly in their teaching. There is a learning curve for us, too. The more we can jump in and try to model with number lines, the more comfortable we'll become and the more comfortable our students will become. We can invite our students to help us and learn together.

Why do some students have so much trouble putting fractions on a number line?

The combination of fragile understanding of fractions and lack of experience using number lines makes the task very challenging. There's really a lot going on at once. Students need to be able to compare fractions, put fractions in order, and also pay attention to the range of the number line and where the fractions fit in that context. Carefully scaffolding the experiences and giving students time to make sense of what they're doing is key. Class discussions, partner talk, justifying answers, and using real-world contexts (rulers, racetracks, etc.) all contribute to the development of the necessary skills and concepts.

Why should we bother teaching students the constant difference strategy?

Students often struggle with subtraction. There are different ways to think about subtraction, and subtraction means different things in different contexts. We often jump to thinking about subtraction as taking away. Another type of subtraction entails thinking about part-whole relationships and figuring out what's missing. For example, *There were 12 candies, Caren ate 7 of them. How many did Rusty eat?*

The third type of subtraction is comparing. In a comparing situation we look at two different quantities and figure out how many more or less one has compared to the other. The constant difference strategy gives a very handy way of thinking about those comparison problems.

Can using number lines really build place-value understanding?

Yes! By carefully choosing the range of the numbers and the points to plot, we can facilitate rich discussions and deep thinking about base ten and the role of tens, groups of tens, and magnitudes of ten in the system. For example, think about this:

Where does 1,000 fit on the following number line? How can you prove it?

1   1,000,000

Tasks like this get students to think about place value and articulate their understanding of the number system.

Guess My Number

Overview

Guess My Number is a simple warm-up activity that gets students thinking and participating. Students always enjoy guessing games, and this one gives them experience with key math practices. In this activity, students think about characteristics of numbers as well as develop strategic thinking. Students guess a secret number from within a range on the basis of being told whether their guesses are greater or less than the number. With logical thinking and a lot of language, they narrow down the possibilities until they guess the secret number. The game (and variations of it—see the extensions that follow) can be played many times over the year.

**NUMBER SENSE
FEATURES**

(see page xv)

Learning Targets

GRADE 3

I can find the missing number in a multiplication or division equation.

I can find patterns in addition and multiplication and explain how they work.

I can multiply any one-digit whole number by a multiple of ten.

GRADE 4

I can estimate how reasonable my answers are based on estimation and mental math.

I can determine whether a whole number from 1 to 100 is a multiple of a given one-digit number.

I can determine whether a given whole number up to 100 is prime or composite.

I can notice and point out different features of number patterns.

GRADE 5

I can study number patterns and figure out their relationships.

I can understand and explain the value of digits in a large number.

I can explain patterns of zeroes when multiplying numbers by powers of ten.

Materials

None

 ## Time

Fifteen minutes to one hour (depending on version)

Teaching Directions at a Glance

1. Choose a secret number.
2. Tell the class the range of numbers the secret number falls within (1 to 10, 50 to 100, 1 to 100, for example).
3. Have individual students guess your secret number; if their guess is incorrect, announce whether your number is greater or less than the number guessed.
4. Continue facilitating discussions until the secret number is guessed.

Extensions

1. Choose a secret number within a wider range (1 to 500, 1 to 1000). Then give one clue about your secret number (ends with 0; odd; the sum of the digits is 10; etc.). Ask students to investigate the possible numbers and play the game as before.
2. Play the game using a fraction or a decimal as the secret number.

Teaching Directions with Classroom Insights

From a Third- and Fourth-Grade Classroom

1. Choose a secret number.

2. Tell the class the range of numbers the secret number falls within.

"I've got a secret number between 1 and 100," I told a group of third and fourth graders.

3. Have individual students guess your secret number; if their guess is incorrect, announce whether your number is greater or less than the number guessed.

"Raise your hand if you want to guess what it is." Almost everyone wanted to guess. I called on Eddie.

"Ten?" he asked.

"My number is greater than 10," I responded.

"How about 50?" suggested Mae.

"My number is greater than 50," I told her.

Since the class was just learning the game, I recorded each child's guess on the board, and drew a "greater than" symbol next to it to provide a hint and help students keep track. I called on Hans next.

"Ninety?" he guessed.

"My number is less than 90," I said as I wrote down *90* with a "less than" symbol.

"Sixty-five?" guessed Carolina.

"My number is greater," I replied.

"Fifty-one?" asked Antoine.

"My number is greater than 51," I told him.

Some of the students were less than pleased with Antoine's guess. Since I had already said that my number was greater than 65 and less than 90, most of the students knew my number was somewhere between 66 and 89. Antoine's guess of 51 was superfluous. There were a few snorts and heavy sighs.

I was concerned that Antoine and several other students would be reluctant to participate if the threat of humiliation loomed. As a teacher, I need to provide the appropriate arena and set the proper tone. I can't let snide comments or belittling remarks slide by. I need to show the class, through my words and actions, that all thinking is valued, and all individuals are respected.

"When we are learning a new game like this," I told the class, "all guesses are fine. It's not OK to make anyone feel bad about his or her guess. We all have the right to think and talk about our ideas without feeling bad. If people are worried about being made fun of, they're not able to do their best thinking. Brain researchers have proved this. Your brain doesn't work as well when you don't feel safe and respected. So, when someone says something, even if you disagree, you need to listen quietly and respond respectfully. Does everyone understand why this is so important?" I paused, making eye contact with everyone. "Now, who else has a guess?"

"Sixty-nine?" Donald ventured.

"My number is greater than 69," I told him as I recorded it on the overhead.

"Seventy-three?" offered Reggie.

"My number is greater than 73 also," I answered. I pointed to the numbers and arrows on the overhead, which were:

$$> 10$$

$$> 50$$

$$< 90$$

$$> 65$$

$$< 51$$

$$> 69$$

$$> 73$$

4. Continue facilitating discussions until the secret number is guessed.

"OK, I'm going to give you a minute or two to talk at your tables about my secret number. Tell each other what you think you know about my number and what number you might want to guess next." After a short

Teaching Tip
Since a safe environment in which to think about new ideas and take risks is critical to developing important math practices, address this head on.

time, I called everyone back to attention. "Can someone tell us something you think you know about my secret number? I'm not asking for a guess right now. I'm asking *about* my number."

"We know it's less than 90 and greater than 73," Abbie said.

"All right," I replied.

"I think it's in the eighties, most likely," added Jack.

"Why do you think that?" I probed.

"Because there's more numbers left in the eighties than in the seventies," Jack explained. "It can't be 70 or 71 or 72 or 73 because we already guessed 73, and you said it's greater. But it can be anywhere in the eighties."

Brief discussions like this help build number sense. Students have time to consider relative quantities and use concepts such as greater than, less than, and in between. They also use logical thinking to try to narrow down the possibilities. Within a few guesses, the students had determined my secret number.

From a Fourth-Grade Classroom: Expanded Version

1. Choose a secret number within a wider range (1 to 500, 1 to 1,000). Then give one clue about your secret number (ends with 0; odd; sum of the digits is 10; etc.). Ask students to investigate the possible numbers and play the game as before.

"Now let's try the super-challenging version of *Guess My Number*," I invited. "This time I'm going to pick a secret number between 1 and 500. That's a lot of numbers. Since there are so many possibilities I'm going to give you one hint. The hint is that my number ends with a 0. Would someone give us an example of a number that's possible?" Many students were ready and willing.

"One hundred," suggested Andi.

"Yes, 100 is possible because 100 ends with 0," I agreed. "How about an impossible number? Can you give an example of a number you can definitely eliminate?"

"Two hundred and sixty-eight," offered Joaquin.

"Right," I said, "because 268 ends with 8 and not 0. So, who wants to guess my number?"

This time I used a T-table to give students a different visual model to use.

> greater than	< less than

"How about 410?" volunteered Linda.

"My number is less." I wrote *410* under "Less than."

"Three hundred and thirty?" asked Eddie.

"Less," I replied.

Ryan tried 150.

"It's greater than 150," I told him.

I said, "I'm getting curious about something. I wonder how many numbers there are between one and 500 that end in 0."

"Fifty," Andi blurted out almost instantly.

I ignored her for the moment. I knew most of the students didn't know the answer, and my goal was for students to generate a variety of ideas

"I'm going to give you a little time to talk at your tables about this question. Don't just come up with an answer. I want to hear how you figured it out. Use paper and pencil so you can prove your idea."

While the students worked on this question, I circulated. I asked Andi about her answer.

"It's simple," she explained, "because every ten numbers end with 0, and 500 divided by 10 is 50."

Andi displayed a deep understanding of the structure of our base ten number system. She had divided 500 into groups of ten to find out how many multiples of ten there were. Those numbers would be the numbers that end in 0. I was impressed with her number sense. She clearly had a grasp of the problem and understood how to use division to find the answer.

Mathematical Practice

Asking students to consider the orderliness of the number system to find an answer encourages them to look for structure (MP7).

Teaching Tip

Introduce an investigation to further challenge students' thinking.

Teaching Tip

Observe students. Ask them to explain their answers.

Although the question itself hadn't been much of a challenge for Andi, I decided to push her by asking her to write about her thinking. "Wow," I said, "that's an interesting way to think about it. Can you write that down and explain it on your paper? That way I'll be able to remember."

Donald also had a "chunking" approach. He counted by tens on his fingers to 100. "Oh, it's 50," he announced upon reaching 100.

"How do you know?" I asked.

He explained. "There's ten numbers that end in zero up to 100, so times that by 5 for 500." Donald understood that the calculation for the first 100 numbers applied to subsequent groups of 100. Once he figured the answer within 100, he just multiplied by 5 to find the result for 500.

I visited some other tables to see how they were doing. My visits uncovered a range of thinking, and I was able to make some important observations about the students' number sense and their understanding of the structure of the base ten number system. Carolina was struggling with 50 times ten. She had been sitting next to Andi and had overheard our earlier conversation. I think she was trying to use Andi's idea, which didn't really make sense to her. I asked her why she chose the numbers 50 and 10, and she couldn't tell me. I was also alarmed to see she was implementing the standard multiplication algorithm to solve the problem. She had written:

$$
\begin{array}{r}
50 \\
\times\,10 \\
\hline
00 \\
50 \\
\hline
500
\end{array}
$$

Fourth graders should be able to multiply any number by 10 with ease. Why did Carolina go to all the trouble using a multistep procedure?

Some other students were counting by tens and writing each number down. Then they were counting up all the numbers on their paper. Others were counting by tens aloud and keeping track with their fingers. While these methods didn't have the elegance of Andi's, the students were at least organizing their work in a systematic way.

I was concerned about the students who were just randomly writing numbers that ended in 0. They didn't seem to be organizing their work or taking control of the problem. Joey, for example, had started by writing the guesses the class had started with: *410, 330, 150*. Then he continued adding to the list: *240, 60, 110 . . .*

Mathematical Practice

Donald found and made use of the practice of repeated reasoning (MP8) to get his answer.

Formative Assessment

Noticing the ways students organize (or don't organize) their work informs our questions and subsequent lesson focus areas.

"So, what's your plan?" I asked, trying to hint that a plan is a good thing.

"I'm writing down numbers that end with 0," he replied.

"How will you know you found them all?" I asked, again hinting at a bigger picture.

"I'll just count them when I'm done."

The conversation appeared to be going nowhere. I decided to let him continue his haphazard quest. I counted on future whole-class discussions and student sharing to help him see other ways to think about the job.

It was time we finished the game. I referred to the three clues we already had. "So, you know my number is less than 410, less than 330, and more than 150. You also know a lot about numbers that end in 0 from all the work you just did. Who would like to take a guess?" I called on Chalisa.

"Two hundred and fifty?" she asked.

"My number is greater."

"Three hundred and fifty?" Tristan tried.

"My number is less."

"Three hundred," guessed Abbie.

"My number is less than 300," I replied.

"It has to be 260, 270, 280, or 290," Donald volunteered.

"Really?" I challenged, hoping to prompt a stronger mathematical argument.

"Yeah," interjected Andi, "because it's between 250 and 300."

"How about 280?" Eddie asked.

"My number is less."

"Two hundred and seventy," Jack stated with authority.

"My number is less," I announced. The students now knew my secret number and hands were flailing frantically. Rather than pick one student to be the hero, I decided to let the whole group answer. "Put your hands down please," I asked as I waited for students to calm down. "I'm going to count to three, and when I say three you will use your indoor voice to say the answer. One. Two. Three."

"Two hundred and sixty!" was the enthusiastic chorus in a slightly louder tone than I had bargained for.

"That's right," I acknowledged. "So now you know how to play the super-challenging version of *Guess My Number*. Maybe next time one of you can think of a secret number and a hint and we can try to guess your number." The students seemed excited by the possibility.

1. Choose a secret number.

I started the activity with fifth graders exactly as I had with the third and fourth graders. We played a quick game with a number between 1 and 100. Then we moved on to a number between 1 and 500 that ended in 0, this time going right into a whole-class discussion about how many possibilities there were rather than taking time for each table to work on it first.

Then I began a third game with the class, in which I planned to pose a lengthier investigation.

2. Tell the class the range of numbers your number falls within.

"OK," I said, "this time I've got a number between 1 and 500. I have one hint for you. The digits in my number add up to 10."

3. Have individual students guess your secret number.

Can anyone give an example of a number it might be?" I called on Courtney.

"Nineteen," she said.

"Yes," I agreed, as I wrote *19* on the board, "19 works because 1 plus 9 equals 10. How about another example?"

"Two hundred and eighty," volunteered Albert.

"Right," I responded, "because 2 plus 8 plus 0 equals 10. How about some other possibilities?"

"Four hundred and forty-two," said Annie.

Seeing that the students understood my rule, I decided to incorporate a bit of estimation. "So, I have this secret number somewhere between 1 and 500. You know there are fewer than 500 possibilities, because the number has to have digits that add up to 10. I wonder how many numbers there are between 1 and 500 that fit my rule. When we played *Guess My Number* earlier, we found that there are 50 numbers between 1 and 500 that end with 0. Do you think there are more or fewer numbers with digits that add up to 10? Let's get some estimates."

"I think there are going to be 50," speculated Bryan.

"I'd say 100," countered Serena.

"Probably 55," Derron estimated

"More like 400!" Vicky jumped in.

I accepted all the estimates without comment. My goal was to get students to think and make predictions. They really didn't have enough information or experience to make an accurate judgment at this point. I just wanted them to start to think—and wonder—so they'd be motivated to investigate.

"OK, here's the plan," I proceeded. "You're going to have some time to work at your tables to investigate this question. Just how many numbers are there between one and 500 whose digits add up to 10? You'll probably want to use paper and pencil to organize your work and keep track of the numbers you find. I also think it will be very helpful for you to work together and talk at your tables. There are probably different ways you can work on this problem, so you can get a lot of ideas from one another. After you've had some time to investigate we'll get back together to discuss your findings and to play *Guess My Number*. Are there any questions about your job right now?" The students were clear on the assignment, so I let them get to work.

The investigation was very rich. The problem was open-ended, and there were opportunities to access it from a variety of approaches. I had a lot of time to circulate throughout the room while the students worked on the problem. My visits to different tables were fascinating and told me a great deal about the students' number sense. Students' various approaches told me a great deal about their developing ideas and how they looked for and used the structure of the number system.

The primary insight I gained was about different views of the number system. Many of the students began randomly listing numbers that fit the rule. Others began systematically, breaking the range of 1 to 500 into smaller, more manageable groups of 100. As they continued working, quite a few of the students noticed patterns emerging. The patterns were powerful tools for organizing their papers and for establishing the total number of possibilities. I talked to the children while they were working, and I also had them write a little about their plan of attack and how they found all the possibilities.

Quite a few groups figured out how to apply the commutative property to identifying numbers that fit the clue. If 19 works, 91 works; if 361 works, so do 163, 136, 316, 613, and 631. As Kate wrote: *Well first we tried to figure out what adds up to ten and we wrote the numbers on a piece of paper and I also got some numbers by putting them backward like: 73–37, 91–19, 64–46.*

Bhavna found a handy use for zero. After she had written all the two-digit possibilities, she added 0 to the end of them for a bunch more. Then she put 0 between the two digits to make more numbers.

Teaching Tip

Ask for an estimate; this begins to engage students in thinking about reasonableness.

Formative Assessment

A great deal of info is gleaned from observing, listening, and asking the occasional open-ended question such as, "What are you working on?" or "What have you discovered?"

Edwin used compensation as a strategy. He noticed that if one digit is decreased by a certain amount while another digit is increased by the same amount, the sum stays the same. He used this discovery to help him organize his work. He wrote: *I started by looking for the smallest number that equals 10. I got 19. After that I went on and got 28. Right when I got 28 a pattern popped in my head. Annie and I started this pattern. It was to make the tens digit number to go up one and the ones digit number to go down one. 19, 28, 37, 46 . . . But after you finish the pattern with the tens you look for the lowest number that equals 10 but in the one hundreds, then 2, 3, and 4 hundreds. But you do the exact pattern as you did in the tens.* Edwin had then systematically listed every number using the method he had described: 19, 28, 37, 46, 55, 64, 73, 82, 91, 109, 118, 127, 136, 145, 154, 163, and so on.

Interestingly, Edwin was not confident that he had found all the numbers. He wrote at the end of his paper: *I don't think you will know all of these numbers through 1 and 500. But if you work on this problem for maybe a while you can find them all.* This taught me a valuable lesson. Even when children find patterns and structure, they need time to realize their usefulness and applicability. The search for and discovery of patterns in the number system is a fundamental building block of number sense and a practice we want to develop in all students. However, the next step is to use these patterns to help solve problems by connecting these patterns to justifiable arguments. The connection does not occur automatically. Children need many opportunities to use and talk about the patterns in order to appreciate their value in problem solving.

Several groups drew columns as their organizing tactic. Some students divided their paper into five categories, 1–100, 100–200, 200–300, 300–400, and 400–500. Then they listed all the possibilities under each column, some doing so randomly, others being more organized.

Albert focused on the digits in each column to provide additional organization. He headed each column with three lines, for three digits. Then he filled in the first digit of each column and listed the possibilities for the second and third digits underneath. (See Figure R4–1.)

I was intrigued by the elegance of this method, which reveals several patterns at once. Looking across the rows shows one pattern. Going down each column shows the pattern Edwin described. The total number of possibilities in each column also appears to have a bell shape and corresponding numerical pattern, 9, 10, 9, 8, 7. I was impressed. Here was an opportunity to consider many properties of the numbers system simultaneously to build number sense.

Teaching Tip

Observe students. What do you learn about students' number sense?

1. I made a graph with my table and buddy. This is how we did it.

0	1	2	3	4		9
1=19	1=09	1=08	1=07	1=06		10
2=28	2=18	2=17	2=16	2=15		9
3=37	3=27	3=26	3=25	3=24		8
4=46	4=36	4=35	4=34	4=33	+	7
5=55	5=45	5=44	5=43	5=42		49
6=64	6=54	6=53	6=52	6=51		
7=73	7=63	7=62	7=61	7=60		
8=82	8=72	8=71	8=70	7		
9=91	9=81	9=80	8			
9	10=90	9				
	10					

2. My table looked at all the combinations and it appears to be no more combinations.

Figure R4–1. Albert used columns to organize his work.

With about ten minutes of math class left, I brought the students back together for a whole class discussion. They were at different points in their investigation, and they would need more time in the next day or two to continue exploring. So, we talked about what they had discovered so far. I closed the session by playing one last game of *Guess My Number*.

"Now you have quite a bit of information about numbers whose digits add up to 10," I told them. "I've seen a lot of incredible work and thinking going on here. I'm going to give you an opportunity to put your knowledge to the test. You can use your papers to help you when we play *Guess My Number*. I have a secret number whose digits add up to 10. Who would like to take a guess?"

Teaching Tip

Bring students back together for a whole-class discussion.

Reflecting on the Lesson

Is this activity really that beneficial?

I feel I get a lot of mileage out of *Guess My Number*. I stop at various points during the game to ask questions that force the students to think about numbers and relative quantities. They construct viable arguments, critique the reasoning of others, and look for structure. The mini-investigations are a great assessment tool. By observing, listening, and questioning the students, I get a window into their number sense and gauge their proficiency with several practice standards.

How can I use this activity in my classroom?

Guess My Number can be used as a warm-up throughout the year. You or your students just need to create new hints or different ranges of numbers. Fractions or decimals can be incorporated as well. Spending time on related investigations is motivating and loaded with learning. Students are able to see relationships and connections between numbers and use these patterns as a potent problem-solving tool.

What are some other hints I can use in games?

There are many options. It's sometimes useful to decide on the range first and then decide how many numbers you want to eliminate with your clue. For example, if the range is 1 through 100 and your clue is that your secret number is odd, half of the numbers in the range are still possibilities. If the range is 1 through 100 and your clue is that your secret number has two identical digits, you've narrowed the possibilities considerably.

You can also ask your students to brainstorm a list of clues they might use if they were leading the game. You'll be impressed with what they come up with and can use their ideas in future games.

Why were you concerned that Carolina used the standard multiplication algorithm for 50×10?

First, both 50 and 10 are landmark numbers. Third and fourth graders need to be familiar with these numbers as quantities and as addends or multiples. These are key elements understanding base ten numbers and operations. Whether or not Carolina had much prior experience in dealing with two-digit multiplication, she should have had ways to think about 50 ten times or 10 fifty times.

Another part of number sense is efficiency. We don't want our students counting on their fingers into the thousands. It's just not efficient. Nor is it efficient to use a multistep paper-and-pencil procedure for a problem that can be solved mentally in a matter of seconds.

Computational Estimation

Overview

Estimation and thinking about the reasonableness of answers are important skills and key to number sense development. In this activity, students are given an arithmetic problem and have several different answers from which to choose. Each "answer" is an approximation, not an exact answer to the problem. Students choose one of the answers they judge as most reasonable, and then they communicate their reasoning to others. Estimating the answer to an arithmetic problem involves the students in performing some initial computations that serve to guide their estimates.

NUMBER SENSE FEATURES

(see page xv)

Learning Targets

GRADE 3

I can determine how reasonable my answers are using mental computation, estimation, and rounding.

GRADE 4

I can determine how reasonable my answers are using mental computation, estimation, and rounding.

GRADE 5

I can use number sense and fractions that I know to estimate the reasonableness of answers to fraction problems.

Materials

None

 **Time**

Ten to fifteen minutes

Teaching Directions at a Glance

1. Write an arithmetic problem on the board.
2. Write several approximate answers to the problem.
3. Ask students to think about which approximate answer is most reasonable.
4. Have students share their reasoning with others.

Teaching Directions with Classroom Insights

From a Third-Grade Classroom

1. Write an arithmetic problem on the board.

To begin the lesson, I wrote the following problem on the board:

$$560 + 1,409$$

2. Write several approximate answers to the problem.

"For this problem, you don't have to figure out the exact answer," I told the class. "I want you to estimate which answer is *closest* to the exact answer." I wrote the following on the board:

about 1,000

about 2,000

Teaching Tip

Computational estimation involves deciding how close you want to get to the actual answer, and this depends on the problem context.

---

about 3,000

about 4,000

3. Ask students to think about which approximate answer is most reasonable.

"Talk with a partner about which one you would choose and explain *why*," I directed. I gave the students some individual think time first before talking with a partner.

As students began chatting in pairs, I circulated around the room, trying to get a sense of the range of thinking. Some students had a hard time making an estimate, and I noticed that they were mentally calculating an accurate answer. Others were able to use their number sense to make estimates by rounding to friendly numbers.

4. Have students share their reasoning with others.

After a minute or so, I called the class back together and took a poll, asking students to raise their hand and let me know which answer they chose. This is one way to "take the number sense temperature" in the room. The polling revealed that three students thought the answer was about 1,000, most thought the answer was about 2,000, and only one person had chosen 4,000 as the *about answer*. Listening to their reasoning would tell me more about their number sense, so I called on several students to explain their thinking.

"It's about 2,000, 'cause I added it up in my head," Israel reported.

"Did anyone think of it without actually finding the exact answer?" I asked.

"Five-hundred sixty plus 400 is about 1,000, plus another 1,000 is close to 2,000," Veronica explained.

"I rounded 1,409 to 1,400," Miguel said. "Then, like Veronica, I knew that 1,400 and 400 is about 2,000."

"Nod your head if you rounded to a friendly number while estimating," I directed. Many students nodded. Rounding when estimating is an important skill to use when engaging in computational estimation, especially if it helps you make an estimate that's close to the actual answer. An important mathematical practice is deciding how close you want to get to the actual answer, and this depends on the problem context.

Teaching Tip

Take a poll, asking students which answer they chose. This is one way to "take the number sense temperature" in the room.

The next problem I posed was *3258 – 300*. As before, I wrote on the board the possible choices:

about 1,000

about 2,000

about 3,000

about 4,000

I wondered if this problem would be too easy for the students since it seemed obvious to me that the answer would be about 3,000. After polling the class, I discovered that no one thought the answer was about 1,000, and most students chose the correct answer, about 3,000. However, there was a big subgroup that thought the answer was about 2,000, and one girl who chose 4,000 (this was the same girl, Maria, who thought the answer to the previous problem was about 4,000). I was curious to hear their arguments.

"I found the exact answer, but I did it mentally," Jason reported. A few students nodded their head in agreement.

"How did you figure the exact answer?" I probed. Jason responded, "I did 8 take away 0 is 8; 5 take away 0 is 5; you can't do 2 take away 3, so I borrowed from the 3. . . ."

Jason was describing the standard subtraction algorithm. While I was impressed that he was able to perform this complicated procedure in his head and get the correct answer, two things concerned me. While explaining the procedure, Jason described the quantities as single digits rather than use language that might indicate an understanding of place value. The second thing that I wondered was why he didn't use estimation. Was he able to estimate? Or was he so used to thinking procedurally that it hadn't occurred to him?

Maria, the girl who got the incorrect answer for both problems, also tried to mentally figure the exact answer using the standard algorithm but made mistakes. The thing that concerned me about her thinking was that she didn't seem bothered by the unreasonable answers that she arrived at.

Some students, those with strong number sense, thought about the answer like this: "Two-hundred fifty-eight and 300 are kind of close, so when you take out those two numbers, you're left with about 3,000." These students strategically used estimation as a mental math tool to arrive at a reasonable result.

Teaching Tip

Provide students with many opportunities to estimate; this helps them develop their number sense.

Finding accurate answers is important during math class and when computing in the world. But estimation is also important. In fact, when we use arithmetic in the world, whether we are at the store, at a restaurant, or figuring the cost of an item that is on sale, we use estimation about half the time. We express numerical answers with a degree of precision appropriate for the problem context. Providing students many opportunities to estimate helps them develop their numbers sense.

Teaching Tip

Estimation can be used as a mental math tool to arrive at a reasonable result.

From a Fourth-Grade Classroom

1. Write an arithmetic problem on the board.

With a group of fourth-grade students, I posed the following problem:

$$365 \times 80$$

2. Write several approximate answers to the problem.

"I'm going to show you some possible answers to 365 times 80," I said. "One of the numbers is close to the answer, but not exact. Which one do you think is closest to the answer?"

I then wrote these numbers on the board:

about 500

about 2,500

about 30,000

about 350,000

about 1,000,000

3. Ask students to think about which approximate answer is most reasonable.

With a show of hands, ten students thought 2,500 was closest, ten students thought 30,000 was closest, four estimated a million, and no one thought 500 could be possible.

4. Have students share their reasoning with others.

I then gave students some time to talk with a partner about the reasonableness of the answers. After a minute, I asked for their ideas.

Belinda said, "I think 500 couldn't be closest to the answer 'cause 365×2 is more than 500."

"What's 365×2?" I asked the class. I gave the students a few seconds to think, then called on Jesus.

"It's 730," he reported. "First I did 5 plus 5 is 10. Then I did 60 plus 60 is 120. Then I doubled 300 and that's 600. I added 120 plus 10 plus 600 and that's 730."

"So, 500 isn't reasonable," I said. "What about some of the other numbers?"

"A million is way too big!" Marco exclaimed. Everyone seemed to agree with his idea.

"What about 2,500?" I asked. "Reasonable or not?"

"No, it's not a good number," Rain said. "If you multiply 300 times 10 you'd get 3,000 'cause 100 times ten is 1,000 so 300 times 10 has to be 3,000 and that's more than 2,500!" Mathematically proficient students calculate accurately and efficiently, and they communicate precisely to others. Rain was able to multiply numbers by 10 and multiples of ten easily in his head—a skill that shows that he is able to use place value when operating with numbers.

Rain's reasoning seemed to make sense to the class. Focusing on *what makes sense* when computing is an important mathematical practice.

"So, all we have left is 30,000 or 350,000," I said. Next, I asked for a show of hands to find out what students' estimates were now that they'd had a chance to hear different ideas. This time, everyone thought that the exact answer was closest to 30,000, except for Henry, who stuck with 2,500.

"To find out the exact answer to 365×80, what can we do?" I asked the class.

"Use a calculator!" Amber suggested.

"Multiply," Anton said. "I already know the answer!"

"OK, we'll have Amber get a calculator and see what she finds out," I said. "Anton, you can explain how you got the answer in the meantime." Considering the appropriate tools to use when computing (in this case, paper and pencil and a calculator—with the other problems, mental math was a useful and appropriate tool) is part of making sound decisions when problem solving.

After Anton explained how he figured the answer using the standard algorithm, we checked his result with Amber. We didn't spend time

Teaching Tip

Ask students, "Does that make sense?" This shifts the teaching focus to number sense development.

Mathematical Practice

Mathematically proficient students calculate accurately and efficiently, and they communicate precisely to others (MP6).

exploring other ways to find the exact answer because the focus of the lesson was on using computational estimation to think about reasonable answers, not exact ones. When Anton and Amber reported the answer (29,200), I ended the lesson by asking the students which number was closest to the exact answer: 30,000 or 350,000.

"It's closest to 30,000," Jalen said. "If you round 29,200 up to the nearest friendly number, it's 30,000. Three-hundred fifty thousand is way too big!"

From a Fifth-Grade Classroom

1. Write an arithmetic problem on the board.

To provide a context for the problem 3.45×25, I told the group of fifth graders the following story:

A girl got paid $3.45 an hour to help her parents around the house. She worked for 25 hours. How much money did she get paid?

2. Write several approximate answers to the problem.

After telling the story, I showed the class the possible *about answers* they could choose from:

about $8.00

about $80.00

about $800.00

about $8,000.00

3. Ask students to think about which approximate answer is most reasonable.

After giving the students some think time and then time to talk with a partner, I took a poll to elicit their ideas. Most students thought the answer was about $80.00, three students thought the answer was about $800.00, and two thought the answer was about $8,000.00.

4. Have students share their reasoning with others.

"Can the answer be all three of those amounts?" I asked. There was a resounding "No!" from the students. "So, who can share what you think and defend your answer?"

"I think she earned about $80.00," Alberto said. "It's because I did 25 times 3 equals 75. So, the answer is around 80."

"How do you know that 25 times 3 is 75?" I asked.

"Twenty-five plus 25 is 50, and another 25 is 75," Alberto explained.

"I just did 3 dollars times 25 and rounded the answer to 80," Mia added.

I noticed that the students who struggled were those who tried to figure an exact answer, while those who used estimation and rounding to friendly numbers were better able to arrive at a reasonable result. Computational estimation is an important skill, especially when taking a multiple-choice test.

Next, I posed this problem:

$$\frac{1}{2} + \frac{2}{3} =$$

I gave the students four options to choose from:

closer to 0

closer to $\frac{1}{2}$

closer to 1

closer to 2

This time I gave the students the following sentence frame to help them explain their thinking in writing:

I think it's closest to _____ because _____.

When the students were finished writing, I held a brief class discussion so that different ideas could be shared. When the lesson was over, I pored through their papers and noticed a range of thinking that revealed a lot about the students' number sense. Of the thirty-two students in the class, twenty thought that the answer was closest to 1 (the correct response). Six thought the answer was closest to $\frac{1}{2}$, and six students thought the best answer was 2.

Several students who got the correct answer used the standard algorithm and found common denominators (see Figure R5–1).

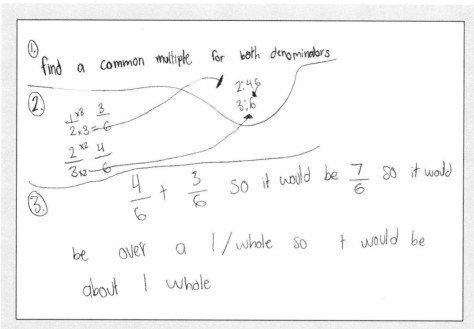

Figure R5–1. Jacob used the standard algorithm to find common denominators.

Many students used visual representations, like circle diagrams (see Figure R5–2) or number lines (see Figure R5–3) to show their understanding of fractions as parts of wholes.

Formative Assessment

Formative assessment includes noticing how students make use of math tools when solving problems.

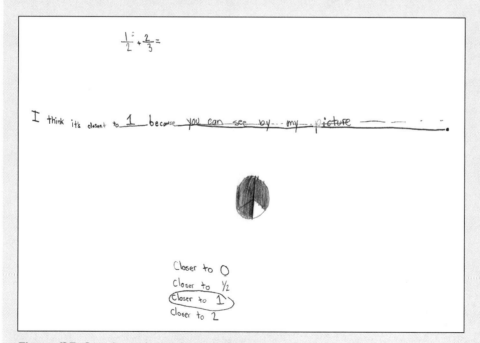

Figure R5–2. Amanda used a circle diagram.

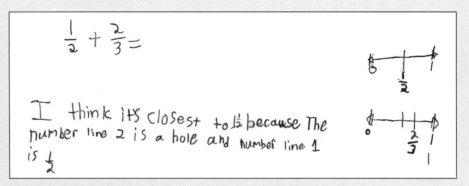

Figure R5–3. Dawn used a number line for her estimate.

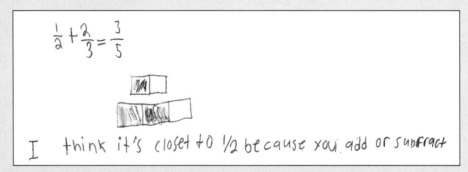

Figure R5–4. Jack revealed his lack of number sense by adding the numerators and adding the denominators.

A few students took out their fraction strips without being prompted, showing that they knew how to make use of math tools to solve problems—an important mathematical practice.

Other students who chose the correct response revealed their number sense through written explanations. For example, Monica explained, "I think it's closest to one whole because $\frac{1}{2}$ is a half, and $\frac{2}{3}$ is almost a whole, so it will maybe be more than a whole but close to a whole." Gustavo wrote, "I think it's closer to one whole because I estimated that $\frac{2}{3}$ is closer to $\frac{1}{2}$ (than it is to one whole) and then I added $\frac{1}{2}$ plus $\frac{1}{2}$ and it was one whole."

Taking the "number sense temperature" in the class by having students explain their reasoning using words and diagrams is a good assessment tool because it can uncover misconceptions. For example, a few students figured that the answer was closest to $\frac{1}{2}$ by adding the numerators together and the denominators together. One student (see Figure R5–4) not only added the numerators and denominators to get $\frac{3}{5}$, but the diagram he drew shows his struggles with visualizing the relative size of fractions.

Reflecting on the Lesson

What are the benefits of computational estimation?

When students estimate the answer to a computation problem, it requires them to round to friendly numbers, consider the place values of the numbers in the equation, and make initial computations that get them "in the ballpark." When estimating, students must think about what happens to numbers when they operate on them. This thinking causes students to predict the magnitude of sums, differences, products, and dividends, and judge the reasonableness of their estimates. All of this mental work serves to develop students' number sense and engage them in the mathematical practices.

How does one help students like Maria (described in the third-grade vignette) develop their number sense?

One unintended consequence of focusing so much on procedural fluency is that it can rob students of their number sense. We have seen this happen in many upper-grade classrooms where teachers use traditional approaches to computation instruction. When students are only asked to memorize procedures taught by the teacher, and then calculate only to find an accurate answer, students' thinking tends to become inflexible. Asking students to estimate the answer to problems *every day* is a good way to begin them on the journey of improving their number sense.

Focus Lessons

Section Overview

A focus lesson is defined by Jennifer Lempp as "a well-planned, whole-group lesson focused on the day's learning target and accessible to all levels of learners" (2017, 121). These activities lend themselves to deeper exploration and connections and are structured in a "launch, explore, summary" format. They include a description of how the teacher introduced the lesson to the class, followed by a description of student work time. Making use of student reflection in summarizing these rich whole-class math tasks is the key to bringing thinking and number sense concepts to the surface. Detailed descriptions of how the teacher summarized the lesson with a deliberate use of questions offers the reader a variety of ways to bring students' number sense to the forefront of the activity.

FOCUS LESSONS

FL–1   Trail Mix   57

FL–2   Numbers and Me   68

FL–3   How Many Beans?   77

FL–4   Fraction Ballparks   88

Trail Mix

Overview

Students benefit from experiences that help them connect abstract ideas about fractions to real-world contexts. In *Trail Mix*, students are given a recipe for trail mix that serves six people. The measurements include fractions, and the students' task is to convert the recipe so that there is enough trail mix to serve everyone in the class.

Learning Targets

GRADE 4

I can understand a fraction $\frac{a}{b}$, with $a > 1$, as a sum of fractions $\frac{1}{b}$.

I can understand the addition and subtraction of fractions as joining and separating parts referring to the same whole.

I can decompose a fraction into a sum of fractions with the same denominator in more than one way and justify my work using models.

I can add and subtract mixed numbers with like denominators.

I can apply my understanding of multiplication to multiply a fraction by whole numbers.

I can solve word problems involving multiplication of a fraction by a whole number.

GRADE 5

I can solve word problems that involve addition and subtraction of fractions.

I can use number sense and fractions that I know to estimate the reasonableness of answers to fraction problems.

NUMBER SENSE FEATURES

(see page xv)

I can use what I know about multiplication to multiply fractions or whole numbers by a fraction.

I can solve real-world problems that involve multiplication of fractions and mixed numbers.

To download the Reproducible, please visit www.mathsolutions.com/mathworkshopessentialsnumbersense.

Materials

Trail mix recipe (see Reproducible 1)

 **Time**

One to two class sessions

Teaching Directions at a Glance

1. Show the students the Trail Mix recipe. (See Reproducible 1.)

2. Review the recipe with the students, clarifying the ingredients and any measurement abbreviations.

3. Have the students determine the number of people in the class.

4. Choose one ingredient for the whole class to convert into the larger amount required for the recipe to serve everyone in the class. Have students work in groups to determine how much of this ingredient would be required.

5. As a class, discuss different approaches.

6. Have students work in groups to convert the rest of the ingredients and rewrite the recipe so that it serves the number of people in the class.

7. Discuss estimation and accuracy with the class.

Extension

Discuss situations in which accuracy is necessary and situations in which estimation is appropriate or even preferable.

Teaching Directions with Classroom Insights

From a Fifth-Grade Classroom

Ms. Barraugh presented her fifth graders with an opportunity to grapple with fractions in a meaningful context. Her goal was to give them a problem in which fractions were integral. She wanted the task to be understandable but didn't want the computational approach(es) to be obvious. She tapped into students' prior knowledge by asking them what they knew about trail mix.

1. Show the students the Trail Mix recipe.

After establishing the basics, Ms. Barraugh showed students the recipe. She asked Annabel to read it aloud.

2. Review the recipe with the students.

When Annabel had finished, Ms. Barraugh pointed to the last line. "It says, serves 6. What does that mean?"

DeSean jumped in. "It serves 6 people. If you make that recipe, it's enough for 6 people to eat."

3. Have the students determine the number of people in the class.

"Well," Ms. Barraugh responded, "what if we want to make enough trail mix for the whole class? How many people would we need to serve?" There was some disagreement here. First, they needed to clarify that they would not include children who were absent. Then they needed to decide whether to include their teacher and their student teacher. The options narrowed down to 26, 27, or 28 people. This discussion was quite useful, because it gave several numbers from which to choose. Ms. Barraugh went with 27.

Not sure how difficult the task would be, and curious to see what approaches the students would use, Ms. Barraugh decided to have everyone work on one ingredient before they split into groups to tackle the entire recipe. She pointed to the recipe and indicated the $\frac{1}{2}$ cup of raisins.

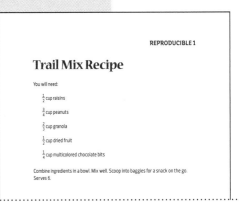

REPRODUCIBLE 1

Trail Mix Recipe

You will need:

$\frac{1}{2}$ cup raisins

$\frac{3}{4}$ cup peanuts

$\frac{2}{3}$ cup granola

$\frac{1}{2}$ cup dried fruit

$\frac{1}{4}$ cup multicolored chocolate bits

Combine ingredients in a bowl. Mix well. Scoop into baggies for a snack on the go. Serves 6.

177

For a downloadable version of R–1, visit www.mathsolutions.com/ mathworkshopessentialsnumbersense.

4. Choose one ingredient for the whole class to convert.

"OK," she said, "a half cup of raisins is what we need to serve 6 people. So, we need to figure out how many cups of raisins we'll need to serve 27. Everyone will need a piece of paper and pencil to work on this problem. It might be helpful to talk with the other people at your table as you think about this problem and work on it, but you each need to end up with a paper that shows your own thinking and work. I'll let you work on this for a while, and then we'll get back together, and some of you will volunteer to show how you thought about it. Make sure your papers have the information you'd need to explain your thinking and how you got your answer."

Ms. Barraugh wrote on the board, *One-half cup raisins serves 6 people. How many cups will you need to serve 27 people?*

"This is the problem you'll be working on at your tables. Are there any questions?"

The students set to work. As Ms. Barraugh circulated, she noticed immediately that there was a range of comfort with the problem. Some students jumped right into the computation or were organizing their work in columns or rows.

Sometimes the biggest challenge is getting a student to articulate the task. Ms. Barraugh directed students to the task at hand through a series of questions, beginning with general ones such as *How's it going?* Then she used students' answers to ask more specific questions such as, *What are you working on right now? What's your plan? What do you already know?* Ralph needed some support.

"So, how's it going?" Ms. Barraugh asked him.

"Good," he answered.

"What are you working on?" she asked.

"We have to do the problem," he answered dutifully.

"What is the problem?"

"Uh, it's a recipe."

"Yes, and what are you supposed to do with the recipe?"

"We have to do the raisins," Ralph continued.

Ms. Barraugh pointed to the question he had written on his paper. "Oh, so the recipe says a half cup of raisins serves six people. So, what are you supposed to do?"

Ralph seemed to brighten. "We have to do the raisins for the whole class."

"How do you think you can find that out?" Ms. Barraugh asked him.

Teaching Tip
Observe the students. What are their comfort levels with the problem?

Mathematical Practice
Understanding the key components of a problem and being able to set up the information and equations in organized ways are key components of the practice, model with mathematics (MP4).

Ralph gave that some thought and began to write on his paper:

$$\frac{1}{2} = 6$$

$$1 = 12$$

$$1\frac{1}{2} = 18$$

"I can do this and keep going until I get to 27," he realized.

"Great," Ms. Barraugh responded. She wondered what would happen when he realized he wouldn't get to 27 exactly. She decided not to confuse him by bringing it up. He understood the problem and was getting into the math, so she moved on. She let him continue working, making a mental note to see how he and other students using the same strategy dealt with reaching 27.

Continuing to circulate, Ms. Barraugh was amazed at the different approaches students took to the problem.

Teaching Tip

Observe the students. What are the different approaches students are taking to solve the problem?

Mathematical Practice

Discussing different approaches provides a context for the practice, reasoning abstractly and quantitatively (MP2).

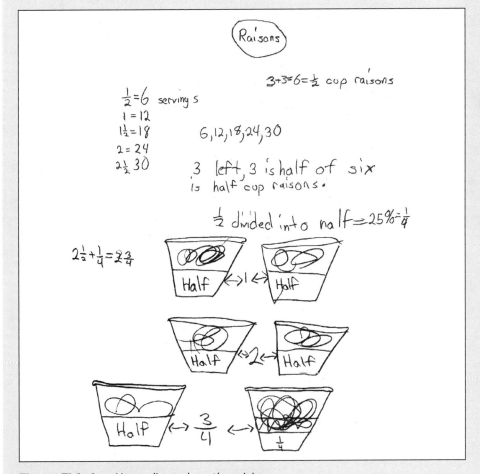

Figure FL1–1. Howard's work on the raisins.

Many students used more than one approach. Howard's paper included words, pictures, fractions, and skip-counting (see Figure FL1–1 on the previous page). Austin and Sean provided a detailed explanation of their thought process.

5. As a class, discuss different approaches.

After the students had been working for about twenty minutes, Ms. Barraugh called them back together as a class. Some were still working on ways to organize their work. Others had found out the quantity of raisins needed for exactly 27 people and had begun to work on converting some of the other ingredients in the recipe. The problem was nicely differentiated in that students had time to work on the first problem and also had challenges and choices when they were ready to move on.

"Who would like to come up and show the class how you approached this raisin question?" Ms. Barraugh asked. To emphasize the communication aspect of this report, she added, "If you're raising your hand, it means you're willing to come to the board and show us your thinking. You'll need to do it two ways. One way is to write on the board to give us an idea of what you did on your paper to help you solve the problem. The other part of the job is to talk to us about what you're writing so we'll understand where your numbers and ideas came from. It's kind of tricky to write and talk at the same time, but it will really help us understand your thinking. Does anyone want to give it a try?"

There were quite a few eager volunteers. Ms. Barraugh called on Frannie, writing her name on at the board. Frannie took her paper from her desk, picked up a marker, and went to work.

"OK," she told us, "a half cup serves six people." She wrote on the board, $\frac{1}{2}$ serves 6. "So, $\frac{1}{2}$ plus $\frac{1}{2}$ equals one whole, and that serves 12." Frannie glanced at her paper and wrote:

$$\frac{1}{2} \quad \frac{1}{2} \quad \frac{1}{2} \quad \frac{1}{2} \quad \frac{1}{2}$$

$$6 \quad 12 \quad 18 \quad 24 \quad 30$$

$$27 = \frac{1}{4}$$

"See," she explained, "I just kept adding $\frac{1}{2}$, and that's 6 more people, and I kept going until I got to 30."

"Can you tell us about the $27 = \frac{1}{4}$ you wrote at the bottom? I don't understand how 27 can be the same as $\frac{1}{4}$," Ms. Barraugh queried.

"Oh," she responded, "It doesn't equal $\frac{1}{4}$. I got to 30, but that was too much since I only want to get to 27. So, I knew $\frac{1}{2}$ of $\frac{1}{2}$ is $\frac{1}{4}$ and that's all I needed to get to 27. So, it's four $\frac{1}{2}$s plus $\frac{1}{4}$ more."

"Now I get it," Ms. Barraugh responded. "Is there a simple way to say four $\frac{1}{2}$s plus $\frac{1}{4}$?"

"It's 2 and $\frac{1}{4}$!" Jamal explained, "because four $\frac{1}{2}$s is the same as 2. We got the same answer but in a different way."

Ramon came up to show how he and his partner modeled the problem:

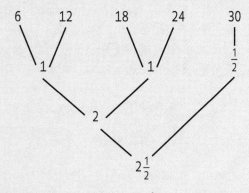

"We counted by sixes, and each 6 is $\frac{1}{2}$ cup," Ramon explained. "So, then we drew lines to add the half cups together and make wholes. Then we added the wholes together, and it takes $2\frac{1}{2}$ cups."

"Actually," Ramon's partner, Veronika, interjected, "$2\frac{1}{2}$ cups is for 30 people but 27 is close to 30 so we just made a little extra."

"Recipes are pretty interesting," Ms. Barraugh noted. "Some cooks like to follow them exactly and other cooks estimate. I guess Ramon and Veronika are estimators."

Ms. Barraugh was glad to open up the estimation option. It would be interesting to see which students went for precise answers and which students got close enough. After a few more students shared their approaches Ms. Barraugh was satisfied students had enough strategies to access the rest of the task.

Mathematical Practice

Pushing for clarity and consensus supports students in the practice, making sense of problems and persevering in solving them (MP1).

6. Have students work in groups to convert the rest of the ingredients.

"Now," Ms. Barraugh announced, "we're going to take some time for you to work at your tables on the rest of these ingredients. Before you get to work, though, it will be really helpful to remember some of the ways different people worked on the raisins. Converting this recipe is not an easy problem at all. You have to work with different fractions and think about ways to increase them, so you'll have enough for 27 people. You have to keep track of a lot of numbers and stay organized as you work. There's a lot of math here. So, before you get back to work, I'm going to remind you of some approaches your classmates used. Maybe after you see some of these ideas, you'll have new ways to think about the problem. They may help you work on some of the fractions that were especially tricky."

Ms. Barraugh wrote on the board, naming the strategies the students had employed to work on the raisins:

Skip-counting by sixes

Skip-counting by fractions

Multiplying

Using a T-table

Using branches

Breaking fractions apart into smaller fractions

Drawing pictures

She told the groups to talk with each other about which strategies they'd like to use for the next part of the problem. Having more than one doorway through which to enter a problem allows for flexible thinking, which is a cornerstone of number sense. If children have only a single procedure for doing a problem, they will invariably be stuck if they forget part of the procedure or encounter a problem that is slightly different from the ones they're familiar with. Flexible thinkers can try a variety of angles and are more likely to find ways to solve the complicated problems encountered in real life. Opening the math class up to multiple approaches also provides access to more students. Not everyone has to think the same way in order to participate.

After the strategy talk Ms. Barraugh assigned each group a specific ingredient to focus on, and the students got to work.

The students easily engaged in the problem. Their previous raisin work and the introductory discussion helped immensely. Several students employed more than one approach when working this time. Apparently, the display of different strategies paid off, giving some of the students multiple ways to attack the problem. Juan used a table and picture to calculate how many tablespoons of sunflowers were needed (see Figure FL1–2).

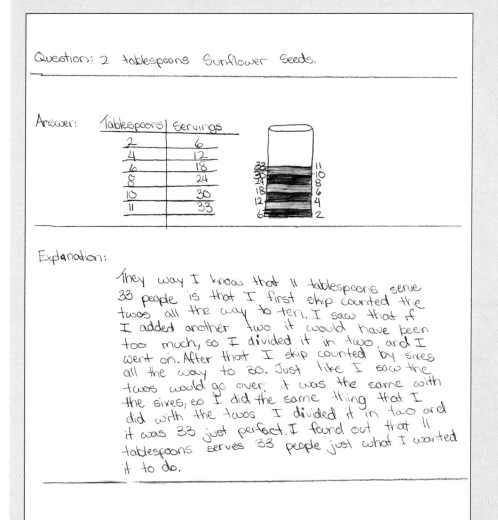

Figure FL1–2. Juan explained his sunflower seed solution.

Raquel used a combination of skip-counting and branching to work on the granola (see Figure FL1–3).

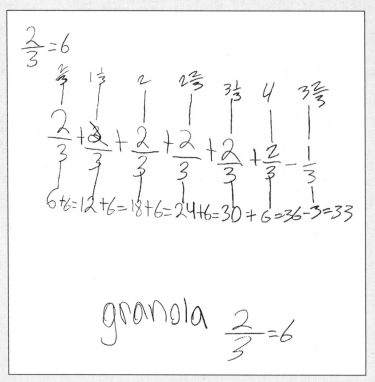

Figure FL1–3. Raquel skip-counted by $\frac{2}{3}$s.

7. Discuss estimation and accuracy with the class.

Ms. Barraugh called the class back together after they had been working about twenty-five minutes. Volunteers reported which ingredient they'd worked on and how much of it they'd need to serve 27 people. Students working on the same ingredient agreed on their answers in most cases. The discussion about estimation and accuracy continued. Students agreed that a close estimate would work in the case of *Trail Mix*. The only controversial ingredient was the $\frac{2}{3}$ cup of granola, so Ms. Barraugh had everyone spend some time working and discussing this problem. Eventually, consensus was reached. The class successfully converted a recipe that served 6 to one that served 27.

Teaching Tip

Hold a whole-class discussion to further encourage students to share and reflect on their thinking.

Reflecting on the Lesson

What is the purpose of this activity?

This activity gives students a lot of experience with fractions in context. They need to think about fractions, compare them, and convert them. They are asked to do quite a bit of computation. An important feature of the activity is the real-world context it gives students for their work with fractions. The numbers and operations used have meaning and purpose. The recipe format is interesting and gets students to think about fractions in a meaningful way, thus helping them develop their number sense.

Is the writing really important?

Writing helps students reflect on their thinking and sharpen their arguments. The class discussions are a prewriting activity. It may seem that by having students talk, they are telling one another the answer. However, with a complex problem, the cross-pollination of ideas doesn't invalidate the written products. Rather, the talk lets individuals focus on what they understand and gives them different ways to think about the problem.

After the discussion, when students write, they pull from what they have internalized. What they have heard through the discussion provides some mental resources, both in terms of ideas and words with which to express the ideas. The ideas clearly expressed on the paper are their own.

How can this activity be used to assess students' number sense?

I have many chances to assess the students throughout the activity. The following questions help me frame my assessment of a student's number sense:

- Which ingredients were relatively easy for students to convert?
- Which were more challenging?
- What computational procedures did students employ to convert the fractions?
- How did students organize their paper to help them see and make sense of the fractions?
- Which students were eager to share their thinking with the whole class?
- Which students tried new approaches after hearing their classmates talk about them?
- Were students able to clearly express their ideas in writing?

FOCUS LESSON

Numbers and Me

Overview

Learning to see the usefulness of numbers is an important part of developing number sense. In *Numbers and Me* students identify some "personal numbers," numbers that describe them or relate to their life in some way. For example, 132 might be the number of a student's house in his street address; 52 might be her height in inches; $11\frac{1}{2}$ might be his age. Then, they play a guessing game with another student in which they try to match the numbers with the things to which the numbers refer. The activity can provide experience with whole numbers, fractions, decimals, and percents.

NUMBER SENSE FEATURES

(see page xv)

Learning Targets

GRADE 3

I understand that fractions are numbers.

I can solve problems involving measurement and data.

GRADE 4

I understand that fractions are numbers.

I can solve problems involving measurement and data.

GRADE 5

I understand that fractions are numbers.

I can solve problems involving measurement and data.

Materials

None

 **Time**

Fifty minutes

Teaching Directions at a Glance

1. Where everyone in the class can see it, write between ten and fifteen numbers that have some significance to your life.

2. Give a clue about each number (for example, "One of these numbers stands for the number of years I've been teaching" or "One of these numbers stands for the number of miles on my car"), and ask the students to decide which number best suits each clue.

3. Have the students write down, on a piece of paper, ten or fifteen numbers that have some significance in their life. On a separate piece of paper, ask them to write a sentence that describes each number. For example:

   2009   This is the year in which I was born.

   3   This is the number of siblings I have.

   $33\frac{1}{3}$ This is the percentage of the day I spend sleeping.

   $\frac{1}{2}$ This is the portion of our class that are girls.

4. Have the students, in pairs, exchange their lists of numbers, then take turns reading their clues (at random) and guessing which number fits which clue.

Teaching Directions with Classroom Insights

From a Fifth-Grade Classroom

Since Pam Long was preparing to teach her students a unit on fractions, decimals, and percentages, I thought *Numbers and Me* would not only be a way to assess students' understanding of these concepts but also give them opportunities to develop their number sense and engage them in the Common Core State Standards for Mathematical Practice.

1. Where everyone in the class can see it, write between ten and fifteen numbers that have some significance in your life.

To begin the activity, I wrote these numbers on the board:

$$33\tfrac{1}{3}$$

$$17$$

$$4$$

$$10$$

$$3$$

$$134{,}155$$

$$0.25$$

$$1954$$

$$60\tfrac{1}{2}$$

$$24$$

2. Give a clue about each number and ask the students to decide which number best suits each clue.

"The numbers on the board are my personal numbers," I told the class. "They all have something to do with my life. They have meaning to me."

"What do they stand for?" Britt asked.

"That's one of the things you'll have to figure out today," I said. "I'll read you clues about the numbers, and see whether you can guess what the numbers stand for. First, I'd like you to read the numbers silently." I waited while the students read the numbers to themselves. Afterward, I had the class practice reading the numbers aloud because I know that this can be tricky for some students, especially when they are reading decimal numbers or very large numbers. For example, when reading decimal numbers like 0.25, many students will say, "Zero point twenty-five," rather than, "Twenty-five hundredths," which describes the place values more accurately.

When we were finished reading the numbers aloud, I continued with the activity. "Now I'll read you clues about my numbers. Listen while I read and see whether you can guess which number I'm describing."

"Here's my first clue," I said. *"This is the year in which I was born."* Many hands shot up, with lots of "oohs" and "ahs." "I believe I'll call on someone who is being silent," I said. As the students quieted down, more hands were raised. I called on Rose.

"It's 1954!" she exclaimed.

"What gave you the clue that it's the year in which I was born?" I asked.

"'Cause all the years in the last century begin with 19," Rose responded. "That gave it away for me."

"OK," I said. "You know the year in which I was born. Can you use that information to help you figure out which number stands for my age?" A few students raised their hands immediately, but I gave the class some time to think about this question. Then I called on Amy.

"I think you are around sixty years old," she said.

"Why do you think that?" I asked.

"Well, I don't think you're 190!" She said with a laugh. "Or 24." Amy understood that most of the numbers on the board didn't make sense. Recognizing the suitability of numbers is part of having number sense. Some numbers are appropriate for some things and not for others.

"Did anyone think about this in a different way?" I asked.

"I did 2015 minus 1954 and got 60," Miguel reported. "The '$\frac{1}{2}$' means half of a year, I think."

Numbers and Me can provide students with opportunities to make sense of and reason about quantities. Thinking about fractions in terms of the number of months out of a year helped Miguel bring meaning to the fraction $\frac{1}{2}$.

Continuing with the next clue, I read, *"This number represents how many brothers and sisters are in my family.* Look at the numbers on the board and think about which ones would fit," I instructed. After giving the students a minute or so to chat at their tables, I called on Tom.

"Well, I think 4 or 3 could be the answer," he speculated. "Maybe 10, but that would be a lot of brothers and sisters!"

"I have three siblings," I stated. "One of the numbers tells what portion of us are boys. Raise your hand if you have an idea."

"It's $\frac{25}{100}$," Miguel said.

"Are you sure?" I asked, pushing for an explanation to check for understanding.

"Yeah," he said, "it's the same as 25 cents and that's like $\frac{1}{4}$ of a dollar."

"Who would like to add to Miguel's comment?"

"It's like 1 out of 4 are boys, or $\frac{1}{4}$," Claire added.

Teaching Tip

Ask for different solution strategies to help students see that there is more than one way to solve problems as well as encourage flexibility in thinking.

Mathematical Practice

Making sense of and reasoning about quantities are important mathematical practices (MP1 and MP2).

"Or 25 percent," Jorge said.

These bits of conversations during the introduction to *Numbers and Me* gave Pam Long, the classroom teacher, insights into what her students knew about fractions, decimals, and percentages. The purpose of the lesson wasn't necessarily to explicitly teach these concepts, but to provide a context for students to think about numbers in a variety of ways and use their number sense to explain their reasoning for their guesses.

"Let's do one more example," I said. "The next number is the age of my car. What do you think?"

"The answer could be three, ten, or 17 years," Jose guessed. No one else raised a hand.

"I'll give you a clue," I told them. "The number that represents the number of miles on my car is 134,155." This clue took awhile for the students to digest and talk about. When I called them back to attention, Tawny was the first to guess.

"How about 3 years?" she said.

"That can't be!" Reba exclaimed. "We have a new car and it has about 10,000 miles on it. You'd have to drive a lot of miles in three years to get over 130,000!" Reasoning about quantities, constructing arguments, and critiquing the ideas of others are important mathematical practices.

"I still think your car is 17 years old," Miguel said.

"The only other numbers that might work are 10, and maybe 24, but that would be real old!" exclaimed Jose.

"My car is 17 years old," I said, revealing the answer.

3. Have the students write down, on a piece of paper, ten or fifteen numbers that have some significance in their life. On a separate piece of paper, ask them to write a sentence that describes each number.

When the students had guessed most of my personal numbers, I explained the directions for the next part of the activity. "You've guessed most of my numbers. Now I'd like you to think of numbers that are special to you. I'd like you to write down as many numbers that you can think of that relate to you, and then write clues about your numbers."

"I don't get the clues part," Tom said, looking confused.

"You're going to write clues about your numbers, just like I gave you," I explained. "For example, if one of my numbers is $60\frac{1}{2}$, my clue might be: 'This number tells how old I am.'" I wrote the number and the clue on the board as an example. "If one of my numbers is 10, I might write: 'This

Teaching Tip

Encourage students to explain their reasoning for their guesses; this helps them develop their number sense.

Mathematical Practice

Constructing arguments and critiquing the ideas of others (MP3) can help students develop their understanding of number sense concepts.

is the number that represents my shoe size.'" Again, I wrote the number and next to it the clue for students to see.

I passed out paper, and the students eagerly began to write down their numbers. One student realized she could ascertain her height using the measuring tape against the wall, and then use that as one of her numbers. Soon, there was a line of students waiting to measure their height.

The students were excited about the activity and worked for the rest of the math period. For homework, I asked students to finish writing their numbers and their clues. I told them it was important that they bring their papers back the next day.

The next day I gave the students about ten minutes at the beginning of math class to look over their numbers and their clues and add any necessary finishing touches. As I walked around the room watching and listening, I was impressed by the variety of ways students thought about numbers. (Figure FL2–1 on the next page is an example of a student's personal numbers and clues.)

Some students wrote about numbers that represented the dates of important events in their lives. Jim wrote down the year that he first got braces. Gwen wrote down the year she got her favorite bike. Tom recorded the year when he'll be 21 years old.

Many students were able to use fractions in meaningful ways. Rich wrote that $\frac{3}{7}$ is the fraction that represents the portion of aunts he has on his mom's side of the family. Leu included fractions that were linked. In his clue about $\frac{12}{34}$, he wrote: *This is the fraction of the class that isn't in the band*. Linda used her height, 5 feet $\frac{8}{12}$ inches, as one of her personal numbers. Students used fractions in a variety of ways, including to describe months in a year, days in a week, states in the union, kids in the class, and inches in a foot.

Some students made important connections between fractions and decimals. Jenny used two numbers to describe her age: 10.5 or $10\frac{1}{2}$. Maryanna, describing 0.5, wrote: When I was 9 years old, I drew 50 pictures and I only liked half of them. Which number means the same as $\frac{1}{2}$?

Students used 100 percent to describe the amount of time our heart beats, how much of the time we breathe, and their score on a test. And they used decimals to describe how much they paid for things, like video games.

Not all the students had an easy time writing their clues. For example, some had difficulty writing large numbers correctly, and I spent time showing them where to place the commas in numbers like 7,545,483. Others struggled with how to describe fractions. When José realized that he didn't have half of a sister, he laughed and rewrote the clue this way: "$\frac{1}{2}$ represents the portion of my brothers and sisters that are sisters." This difficulty with describing what the numbers meant was common and made me realize how important the role of language is in math class. (See Figure FL2–1.)

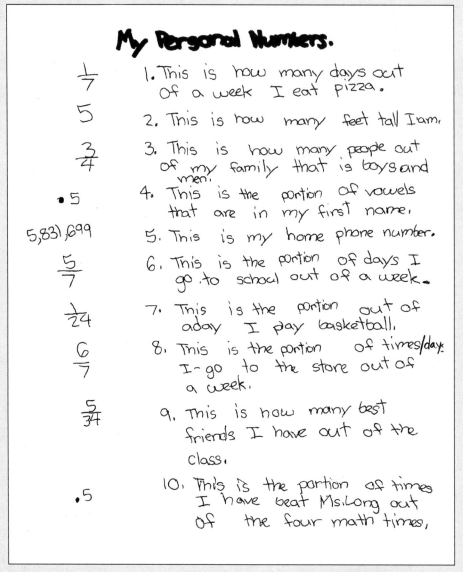

My Personal Numbers.

$\frac{1}{7}$ 1. This is how many days out of a week I eat pizza.

5   2. This is how many feet tall I am.

$\frac{3}{4}$ 3. This is how many people out of my family that is boys and men.

•5   4. This is the portion of vowels that are in my first name.

5,831,699   5. This is my home phone number.

$\frac{5}{7}$ 6. This is the portion of days I go to school out of a week.

$\frac{1}{24}$ 7. This is the portion out of a day I play basketball.

$\frac{6}{7}$ 8. This is the portion of times/days I go to the store out of a week.

$\frac{5}{34}$ 9. This is how many best friends I have out of the class.

.5   10. This is the portion of times I have beat Ms. Long out of the four math times.

Figure FL2–1. José used mostly fractions to describe things in his life.

4. Have the students, in pairs, exchange their lists of numbers, then take turns reading their clues (at random) and guessing which number fits which clue.

Reflecting on the Lesson

How will this activity help my students develop number sense?

This activity helps students develop their number sense in several important ways. First, it prompts them to think about the reasonableness of a number in a given situation. Students have to consider a variety of numbers in the context of any given clue: they need to think logically and eliminate numbers that don't fit. Next, the activity also provides a larger context for thinking about numbers, one outside the confines of the classroom. Students see that numbers are meaningful and can be used in a variety of ways, including to quantify, to label, to measure, and to locate. Providing a meaningful context helps students understand what numbers mean, whether they're fractions, decimals, percents, or whole numbers.

What is the primary focus of the activity?

The primary focus is for students to think about numbers and relate them to familiar contexts. Number sense is characterized by the ability to make sense of numerical situations. We also want students to see that numbers that are appropriate for some situations may be inappropriate for others. Often, students are only asked to think about numbers in very specific ways—when doing arithmetic operations, for example. We want to give students an opportunity to have a conversation that will give them a "feel" for numbers.

During the activity, we deliberately keep the focus on discussing the meaning of numbers in context and away from teaching specific rules or procedures. Context diverts children from rules and procedures and encourages them to explore ideas in a more open and informal manner. In this activity, we want students to see that fractions have meaning in the real world. Understanding what numbers mean is at the heart of number sense. The Common Core State Standards for Mathematical Practices call for a balance between procedural fluency and conceptual understanding. Lessons like *Numbers and Me* help students build conceptual understanding of numbers.

What are some ways to assess an individual student's number sense?

There are many ways to assess whether a student has a well-developed sense of numbers. In this activity, we keep several questions in mind as we observe the students in action:

- What numbers do students choose for their number clues? Do they choose only whole numbers? Are they comfortable with decimals? percentages? fractions?
- What's the range of numbers the students choose? Do they choose only small numbers or do they use large numbers as well?
- When students are playing the game with a partner, are they guessing suitable numbers after being given a clue? Do the students know whether other numbers fit the clue?
- When writing their clues, are students using a variety of contexts for their numbers? What kinds of contexts are they using?

How can I do the activity without using personal numbers?

The activity can easily be adapted for other areas of the math curriculum. In a unit on measurement, for example, students can write number clues about different objects in the classroom: *size 9, 6 feet tall, 3 feet by 5 feet, 11 pounds*, and so on. The activity can also be used in other subject areas. Students in a fifth-grade social studies class, for example, might use important numbers in U.S. history, such as 13 (for the 13 original colonies) and 1776.

How Many Beans?

Overview

A jar, a scoop, and a quantity of beans are common materials that students can use in estimating activities. In the whole-class activity *How Many Beans?* students first estimate how many beans a jar will hold. Then they determine the number of beans in a scoop and begin filling the jar with scoops of beans, adjusting their estimates with each scoop they add. Repeating this activity with different-size jars, scoops, and beans can provide further experiences with estimation.

Learning Targets

GRADE 3

I can determine how reasonable my answers are using mental computation, estimation, and rounding.

GRADE 4

I can determine how reasonable my answers are using mental computation, estimation, and rounding.

GRADE 5

I can determine how reasonable my answers are using mental computation, estimation, and rounding.

NUMBER SENSE FEATURES

(see page xv)

Materials

A jar
A scoop or a large spoon
A bag of dried beans

 Time

Forty-five minutes

Teaching Directions at a Glance

1. Show the students an empty jar, a scoop, and a bag of beans. Ask students to estimate how many beans the jar will hold and record their estimates.

2. Have the students determine how many beans a typical scoop holds.

3. Put a few scoops of beans in the jar and ask the students to mentally calculate approximately how many beans are in the jar and to explain their thinking.

4. Ask students to reestimate how many beans the jar will hold.

5. Repeat Steps 3 and 4 until the jar is full.

6. Have students count the beans in the jar to find an exact total; ask them to compare this total with their final estimate.

Teaching Directions with Classroom Insights

From a Fourth- and Fifth-Grade Classroom

1. Show the students an empty jar, a scoop, and a bag of beans. Ask students to estimate how many beans the jar will hold and record their estimates.

"About how many beans do you think will fit in this jar?" I asked Christina Stamford's fourth and fifth graders, holding up an empty jar in one hand and a bag of kidney beans in the other.

"You mean to the very top?" Tammy asked.

"Yes," I replied. "To the very top."

"*Exactly* how many beans or can we say *about* how many beans?" Joe asked. Tammy and Joe's questions were examples of proficient students trying to identify a problem's givens and constraints, an important math practice.

"I'm interested in having you think about making a reasonable estimate, not in finding an exact answer," I said. "Finding the exact number of beans isn't important for this activity, or really for any reason. I can't think of a situation when knowing an exact number of beans is useful. But thinking about estimates is useful for your math learning."

I walked around the room, giving students a closer look at the jar and the beans. Students began to whisper their estimates to one another.

"Raise your hand if you want to share your estimate with us," I said. A lot of hands shot up. I began calling on students and recording their estimates on the board. Their estimates ranged from 100 to 1,000 beans, with most estimates between 100 and 350. Only one student, Mark, thought the jar would hold 1,000 beans.

Mathematical Practice

Identifying a problem's givens and constraints is an important math practice (MP1).

2. Have the students determine how many beans a typical scoop holds.

When I'd written all their estimates on the board, I took a small scoop, filled it with beans, and held up the scoop of beans and the jar so everyone could see. "I have a scoopful of beans and this empty jar," I said. "Can someone think of a way we could find out about how many beans will fit in the jar?" I asked this question in order to give the students a chance to plan a solution pathway, rather than just tell them.

"That's easy," Jaz said. "All you have to do is find out how many beans there are in the scoop, then fill the jar and count the scoops." Students nodded their agreement.

"You could just pour scoops into the jar and we could count by the number of beans in the scoop," added Jill.

"What if the jar held 10 scoops of beans?" I asked. "How would that help us know about how many beans there are?" Again, hands shot up. This group seemed to find this to be an obvious question, but I wanted them to think and talk about what we needed to do in order to solve the problem.

"You have to use multiplication to figure that out," said Joe.

"What would you multiply?" I asked.

Students benefit
from thinking and
talking about what
they need to do to
solve a problem.

"You multiply the number of beans in each scoop times the number of scoops," he replied. Everyone seemed to be listening intently to Joe's explanation. (I often tell my students that it's important to be quiet while someone is explaining his or her thinking, but that it's more important to try to understand what's being said.)

"Does that make sense? Joe said that you need to multiply the number of beans by the number of scoops," I clarified. "So, we need to know about how many beans there are in a scoop." I poured a scoopful of beans on the table where Megan was sitting and asked her to quickly count them. She reported there were 29 beans in the scoop. "Do you think every scoop will have the same number of beans in it?" I asked the class. They responded with a chorus of nos. "Why not?" I asked.

"Because some of the beans are probably big and some are little," Crystal explained. "They're different sizes."

"That's right, Crystal," I concurred. "Each scoop might have a slightly different number of beans."

I knew that taking one sample wasn't going to give us the best number for a typical scoopful. I could have given each pair of students in the class a scoopful of beans and had them count them, after which we could have taken an average to arrive at a typical number of beans per scoop. But to keep the focus of the activity on estimation and mental calculation, I decided to collect only one sample.

"I'm going to put scoopfuls of beans in the jar, and you're going to keep track of the number of beans," I continued. "But could we use another number of beans per scoop instead of 29, to make it easier for us to count?" I wanted the students to think about using friendly numbers in problem situations. Rounding the number of beans made sense, since they'd already realized that each scoop wouldn't have the same number of beans.

"How about 30 beans?" suggested Barbara. "That's only one more, and 30 is easier to count by than 29."

3. Put a few scoops of beans in the jar and ask the students to mentally calculate approximately how many beans are in the jar and to explain their thinking.

"Watch carefully while I scoop beans into the jar," I instructed. "I want you to count the running total of beans out loud. Let's also keep track of how many scoops fill the jar." I asked for a volunteer to record the numbers on the board. Then I took a level scoopful of beans and poured them into the jar. I held the jar up high, so everyone could see.

"Thirty!" the class chanted.

I poured three more scoopfuls of beans into the jar.

"Sixty, 90, 120," students counted. Skip-counting by thirties was easy for these fourth and fifth graders. Everyone was engaged, their eyes fixed on the jar, which now held a layer of beans at the bottom.

4. Ask students to reestimate how many beans the jar will hold.

"That's 4 scoops," I said. "Raise your hand if you'd like to change your estimate." About a fourth of the class raised their hand. "It's okay to change your estimate," I assured them. "In fact, I want you to think about your estimate every time you see me pour more scoops into the jar. Think about what your new estimate might be and why." Asking students to think about their new estimate based on what's already in the jar is important. Having a reference, or benchmark, to guide their thinking helps them produce reasonable estimates. Mathematically proficient students monitor and evaluate their progress and change course during problem solving, when necessary. Teacher questions can prompt this type of practice by students.

Mathematical Practice

Mathematically proficient students monitor and evaluate their progress and change course during problem solving, when necessary (MP1).

5. Repeat Steps 3 and 4 until the jar is full.

I poured 3 more scoops of beans into the jar as students counted: "One hundred fifty, 180, 210." After seven scoops, the jar was over a third of the way filled. "About how many beans do you think the jar will hold now?" I asked, waiting until many hands were raised.

Wait time, as it's commonly referred to, is critical during class discussions. It allows students to formulate thoughtful ideas rather than quick guesses. It also helps you include the children who are not fast thinkers or strong personalities. Waiting isn't easy for me; my impulse is to call on the first student whose hand goes up. Although this moves the discussion along, it doesn't serve students' thinking. Children need time to think about numbers in order to develop their number sense.

"I think there will be about 500 beans," said Jill.

"Because . . . ," I prompted, helping Jill justify her reasoning.

"Because there's already 210 beans and there's room for a lot more," she explained.

"Other ideas?" I asked.

Teaching Tip

Provide wait time to allow students to formulate thoughtful ideas rather than make quick guesses.

Teaching Tip

Help students
justify their
reasoning by using
prompts and
questions.

"I think there's gonna be 500, too, because the jar isn't half full yet and 400 is twice 200, so it's gonna be more than 400," reported Nick. Using scoops of beans as a context for mental computation helps students make sense of quantities and their relationships in problem situations.

I continued scooping beans into the jar. "Two hundred forty, 270, 300," students chanted. I held up the jar and walked slowly through the room so that the students sitting in the back could get a closer look.

I returned to the front of the room and said, "We have ten scoops of beans in the jar. That's about 300 beans so far. Now how many beans do you think the jar will hold?"

"About two times 300," offered Simon.

"Why do you think that?" I asked.

"Because the jar's about half full and 300 times 2 is 600," he explained. Estimation often involves mental computation as a preliminary step. Lots of students nodded their head in agreement.

"I think about 600, too," said Jean. "But I thought about it different. We have 10 scoops now, so we'll have about 20 scoops when we're done, and ten times 30 is 300, so 20 times 30 is 600." Jean's quantitative reasoning is the kind of thinking we want students engaged in.

Reba waved her hand vigorously after Jean's comment. "I don't think it'll be twice as much, because the jar is bigger at the top." She had noticed that the jar was slanted, so that its circumference continued to get bigger bottom to top. Reba's observation caused a stir, and a lot of hands were raised.

"I think the jar will hold more than 600 because of what Reba said," Josh agreed.

Josh's comment gave me an idea for a question that would prompt all the students to rethink their estimate. "Raise your hand if you think the jar will hold more than 600 beans," I said. Most students did. "Raise your hand if you think the jar will hold about 1,000 beans," I continued. This time no one raised a hand, not even Mark, who had originally thought the jar would hold 1,000 beans.

"I think it'll hold somewhere between 600 and 800," said Tammy. Estimating gives students a chance to compare quantities and think about number relationships by using relative terms like *between*, *about*, *near*, *close*, and so on.

"Let's keep scooping so that we can find out about how many beans the jar will hold," I said. I put scoop after scoop into the jar as the students counted: "Three hundred thirty, 360, 390, 420, 450, 480, 510, 540,

570." The entire class was focused on the jar, the scoop, and the beans. Comments were flying.

"It's almost full!" cried Jaime.

"It's going to be over 600!" Reba exclaimed.

"It's gonna hold another 30, probably 60!" Mark added.

I continued to scoop beans until the jar was full—22 scoopfuls, or about 660 beans.

After an initial moment of excitement, the room grew quiet. Then Reba raised her hand. "Are there really 660 beans in the jar?" she asked. Up until now, the activity had been about estimation, not exact answers. Reba was shifting the focus, and if she hadn't asked the question, I would have.

I handed the question back to her. "What do *you* think?"

"Well, I don't think so, because every scoop didn't really have exactly 30 beans," she said.

"Do you want to find out how many beans there really are in the jar?" I asked the class.

"Not really," Juan said. "They're only beans. Let's just stick with the estimate." Other students, however, were curious about how close our estimate was.

I used Juan's comment to propel the activity in another direction. In order to build number sense, children need to have opportunities to estimate and opportunities to be precise. They also need experience making decisions about how precise an answer needs to be, and this depends on the problem's context. "Maybe some of you aren't interested in finding the exact number of beans in the jar," I said, "but when *would* it be important to figure an exact answer? And when is an estimate good enough?" I allowed the students some time to think and talk about different situations that required estimates and exact answers, then I asked them to report their ideas. "So, what do you think?"

"Well, I agree with Juan that with the beans I would only have to have an estimate," said Simon. "But when I go to the store and buy something, I want to get the exact change when I pay for something."

"If the beans were like quarters or something, and we wanted to share them, I'd want to know exactly how many," said Megan.

"You need to be exact when you do your taxes, that's what my mom says," added Crystal.

"Sometimes my dad just estimates when he cooks," said Juan. "He just throws in a little of this and a little of that."

"So, what about the beans?" I asked. "If an estimate is okay, how close is close enough? Our estimate is 660. How close do we have to be

Teaching Tip

In order to build their number sense, give students opportunities to estimate and opportunities to be precise.

Teaching Tip

Use real-world contexts to bring meaning to quantities.

in order for our estimate to be reasonable?" This question stumped the class. I don't think they really understood what I was asking.

"If there are really 2,000 beans in the jar, would you be satisfied with our estimate of 660?" I asked.

"No!" they responded.

"Why not?" I asked.

"Because that's way off," said Jill. "It should be closer than that."

"What if the actual number of beans is 700? Would 660 be close enough then?" I asked. Students nodded and seemed content with this amount of difference.

"I'd be happy with anything that was about one hundred away," said Jill.

6. Have students count the beans in the jar to find an exact total; ask them to compare this total with their final estimate.

"Let's count them and see how many beans there are!" Nick piped up. So that we could accomplish Nick's suggestion quickly, I poured some beans on each table for partners to count. As pairs finished counting, I wrote the totals on the board. Together, we calculated that the jar held exactly 702 beans.

Reflecting on the Lesson

What is the purpose of this activity?

This activity has two important aspects. One is that it is an experience designed to improve estimation skills. During the activity, students are continually provided with benchmarks that help guide their thinking and improve their estimates. For example, students were able to readjust their estimates several times, first after the jar held 4 scoops, then after it held 7 scoops, again after 10 scoops, and finally when it was full, at 20 scoops. Along the way, students readjusted their thinking based on the new mathematical information made available. Using these benchmarks allowed students to monitor and evaluate their progress and change course if necessary.

The other important aspect of the activity involves mental computation. To arrive at a reasonable estimate, students are required to calculate mentally as a preliminary step. For example, when the jar was about half full of 10 scoops and about 300 beans, Simon offered this line of reasoning: "[There are] about 2 times 300, because the jar's about half full, and 300 times 2 is 600." This is the kind of quantitative reasoning that signals good number sense.

Is it important to use beans in this activity?

There's nothing magical about using beans; cubes or pennies work just as well. What's important is that these concrete objects provide a context for thinking about a problem. A math problem with a context is more meaningful for students and gives them a purpose for computing and estimating. Providing a context (a jar, a scoop, and beans) allows students to represent the quantity of beans using equations.

Whatever manipulatives you use helps students view numbers as quantities and establish benchmarks from which to make estimates. For example, the first time I did this activity, when the jar was a little over a third of the way filled, it held about 210 beans. Students used this reference to guide their thinking when estimating how many beans the jar could hold. In other words, the beans helped them think mathematically rather than make a wild guess. The manipulatives also give students a way to verify whether their final estimate is reasonable or not. Once they arrive at a final estimate, they are able to count the items and compare the total number to their estimate.

You mention that one sample isn't sufficient to find the best number of beans for a typical scoop. Can you say more about this?

As teachers, we make decisions based on the goals of the activity and the amount of time we have to teach it. In *How Many Beans?* one of my goals is for students to gain experience estimating and calculating mentally. The focus isn't on collecting and organizing data or learning about averaging. Finding the typical number of beans in a scoop is certainly important here, but finding the *best* number takes too much time.

How can I assess students' number sense in this activity?

During the activity, I keep these questions in mind in order to assess students' number sense:

- Are students' estimates reasonable or not?
- Are students' estimates based on some mathematical reasoning?
- Are students improving their estimation skills through experience? Do they make use of benchmarks in order to improve on their estimates?
- How large or small a quantity of beans are students comfortable with?
- Are students able to calculate mentally? What strategies do they use? Do they use mental calculation to assist them in making better estimates?

- Do students use fractions when making estimates? Do they seem to understand the fractions they're using?

You can also assess students' number sense by having them write about their thinking. This is what I did when I went into Christina's class for another estimation lesson. For this lesson, the scoop held about 40 beans. At one point in the lesson, we had put 20 scoops into the jar, and 20 scoops equaled about 800 beans (the jar was a little more than half full). I stopped the lesson and gave the students the following writing prompts to assess their thinking (see Figures FL3–1, FL3–2, and FL3–3):

20 scoops are about 800 beans because _____.

Now I think the jar will hold _____ beans because _____.

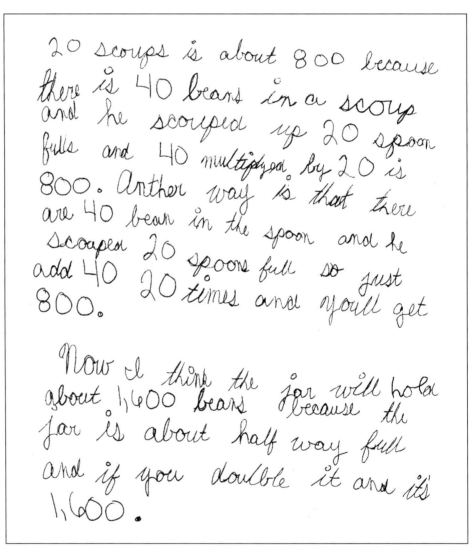

Figure FL3–1. Megan explained why 20 scoops is about 800 beans.

20 scoops is about 800 because

10 scoops is about 400 + 10 more scoops is 800.

Now I think the jar will hold about 1,200

beans because the jar is a little more than

have way fool, and it cant hold much more.

Figure FL3–2. Orlando thought the jar would hold 1,200 beans because at two-thirds full the jar holds about 800 beans.

20 scoups is about 800

because
$$\begin{array}{r} 20 \\ \times 40 \\ \hline 00 \\ *800 \\ \hline \end{array}$$

and 20 scoups

times 40 is *800 beans.

Now I think the jar will hold 10,000 because it is 3 qaters in till it's all the way up.

Figure FL3–3. Tasha's estimate for the number of beans is unreasonable.

Fraction Ballparks

Overview

While many estimation activities involve students in thinking about whole numbers, *Fraction Ballparks* engages them in thinking about fractions, a key content area in upper-elementary grades. Students first think about real-world examples for various quantities—less than $\frac{1}{4}$, about $\frac{2}{3}$, close to $\frac{7}{8}$, and so on. They then analyze other real-world examples and apply them to the students in their class.

The *Ballparks* activity structure works well with whole numbers, too. In a third-grade class, teachers might use *close to 100, close to 250, close to 500,* and *close to 1,000.* Ballparks can even be ranges of numbers: *1–50, 51–100,* and so on. As students develop solid foundations and clear pictures of what numbers look like, teachers can use more challenging quantifications.

NUMBER SENSE FEATURES

(see page xv)

Learning Targets

GRADE 3 (WHOLE-NUMBER VARIATION)

I can solve two-step word problems that involve addition, subtraction, multiplication, and division.

I can use mental math to decide if the answer is reasonable.

I can use place value to help me round to the nearest 10 or 100.

GRADE 4

I can use mental math, estimation, and rounding to determine how reasonable my answer is.

I can recognize and generate equivalent fractions.

I can compare fractions with different numerators and denominators by comparing them to benchmark fractions.

I can compare fractions using symbols and models.

GRADE 5

I can solve word problems that involve adding and subtracting fractions.

I can estimate the reasonableness of my answers to fraction problems.

Materials

6 pieces of chart paper, each bearing a label:

$$\text{less than } \frac{1}{4}$$

$$\text{about } \frac{1}{3}$$

$$\text{about } \frac{1}{2}$$

$$\text{about } \frac{2}{3}$$

$$\text{about } \frac{3}{4}$$

$$\text{more than } \frac{7}{8}$$

$8\frac{1}{2}$-by-11-inch sheets of paper labeled with different categories (*are only children, have an older brother, had cereal for breakfast,* etc.)

🕐 Time

Forty-five minutes to one hour

Teaching Directions at a Glance

1. Post the sheets of labeled chart paper, in random order, along the front of the room. Define *ballpark.*

2. Have students generate real-life examples of things that would fit under each quantitative heading.

(continued)

3. Present the students with one of your predetermined categories, have them apply it to the students in the class, and ask them to discuss which quantitative heading it fits under and why.

4. Have students do a quick-write about the category.

5. Show the students the rest of your predetermined categories and have them choose another one to explore and write about.

Teaching Directions with Classroom Insights

From a Fifth-Grade Classroom

1. Post the sheets of labeled chart paper, in random order, along the front of the room. Define *ballpark*.

I posted six pieces of chart paper at the front of a fifth-grade classroom, labeled and arranged in this order: *about* $\frac{1}{2}$, *less than* $\frac{1}{4}$, *about* $\frac{1}{3}$, *more than* $\frac{7}{8}$, *about* $\frac{3}{4}$, *about* $\frac{2}{3}$.

"OK," I said to the class, "the posters taped on the wall represent ballparks. Does anyone know what I mean by *ballpark* here?"

"It means *about*,'" Rafael responded.

"Like estimating a number," Nida added.

"Right," I replied. "When I say, 'Give me the ballpark price of that car' or 'What's the ballpark number of kids in this school?' I want to know about how much or about how many. The papers I put up on the board are some ballpark fractions for you to think about.

2. Have students generate real-life examples of things that would fit under each quantitative heading.

Can anyone think of a real-life example of something that fits under one of these categories?"

Sensing some hesitation and seeing quite a few puzzled looks, I narrowed the question. Pointing to the sheet that said *about* $\frac{1}{2}$, I said, "Think about this school. Is there something here at school that you can say is about $\frac{1}{2}$?"

Mick brightened. "There's about one-half girls and one-half boys at school."

"Not in our class," countered Tanetta. "We have way more girls in our class."

"But what about the whole school?" I asked. "There are probably some classes that have more girls and others that have more boys. Do you think overall in the whole school it's safe to say about half of the students are girls and half are boys?"

"I guess *about* a half," Tanetta conceded.

"What are some other real-life examples that might fit under one of these ballparks?" I continued.

"I know," offered Guillermo. "More than $\frac{7}{8}$ of the kids play soccer at recess."

While I personally had not spent a lot of time with the kids at recess, the nods of assent from the rest of the class indicated that soccer was indeed a popular activity.

Sariah piggybacked on Guillermo's recess estimate. "Less than $\frac{1}{4}$ play Double Dutch at recess," she volunteered.

"And less than $\frac{1}{4}$ eat lunch at the picnic tables," Leu added. "Almost everyone likes to eat in the cafeteria."

Now that the students understood my original question, I widened the field. "Okay, so now think about examples outside of school," I told the class. "Take a few minutes to talk to your neighbors and see whether you can come up with an example or two for each of the ballparks." I let the students talk with one another at their tables and then called them back to attention.

"Who has an example?" I asked.

Kathy raised her hand. "We think about $\frac{1}{2}$ of the people in the United States like pizza," she reported. Quoch's table had talked about pets. "About $\frac{1}{3}$ of the families in this school have dogs," he predicted.

Brenda had a pie comparison. "When you cut up a pie for people to eat, each piece is less than $\frac{1}{4}$ of the pie."

I made sure the students provided at least one example for each ballpark posted, writing their ideas on the corresponding sheet. Then I continued to explain the activity.

3. Present the students with one of your predetermined categories, have them apply it to the students in the class, and ask them to discuss which quantitative heading it fits under and why.

"Here's what's going to happen now. We're going to flip it around. I have some pieces of paper with me on which I've written some categories of things. When I hold up the first category, your job is to think about which of these ballpark numbers the category belongs under."

I held up the first piece of paper, which said, *Boys in the room.*

"Now take a few minutes to talk at your tables about where you think *Boys in the room* fits. And when you talk to each other, make sure you're explaining your thinking, not just telling answers. It's the thinking that's really important here and your reasons for your answers."

I deliberately gave these instructions to ensure that important practices would be at the forefront of the students' work and thinking.

I circulated among the tables as the students discussed where *Boys in the room* might fit. All students began talking eagerly and animatedly, but the effectiveness of their discussions varied greatly. Some students fixated on determining the exact number of boys in the class. They counted and recounted the boys present and identified who was missing. Other students quickly agreed that there were fewer boys than girls in the class, eliminated the *about* $\frac{1}{2}$ option, and concentrated on identifying which of the other categories could be used to quantify the number of boys. I called the class back together and asked students to share their ideas.

Alicia began. "I think it's about $\frac{1}{3}$," she announced, "because there aren't as many boys as girls in the class."

"It could be less than $\frac{1}{4}$," countered Alberto.

"Why do you think that?" I asked him.

"There are 9 boys in the class and 17 or 18 girls," he explained.

Although the numbers didn't support his idea, I decided not to push him at this point but to check in with him later. My goal was to explore different ways to think about the question. Eventually, I'd push Alberto to prove his conjecture using the language and data of mathematics. For now I was satisfied with an open discussion and different ideas floating around (even incorrect ones).

I looked around the room and counted the boys. "I just counted 9 boys in the room. How many people are there in the room altogether?" After a moment of counting, we determined that there were 26 people in the room, including the students and the teachers.

4. Have students do a quick-write about the category.

"OK," I said, "so there are 9 boys out of 26 people altogether." I wrote $\frac{9}{26}$ on the board. "I'm going to give you a few minutes for a quick-write. Each of you will write your ideas about which ballpark you think $\frac{9}{26}$ is closest to. Don't just write an answer. Explain why you chose the ballpark you did."

"Can we make a picture to show it?" Mary asked.

"Sure, a picture or diagram is a great way to show an idea," I replied, "but you'll also need to write some words to explain your picture and how it helps you."

I gave the class about ten minutes for their quick-write. The papers provided rich assessment information. I could see what strategies students used, how well they explained their thinking, and whether their result was reasonable. Alicia drew a pie graph divided into 26 sections and shaded 9 of them, showing that it was close to $\frac{1}{3}$. She also rounded 26 to 27 in her written explanation (see Figure FL4–1 on the next page). Even the papers of the students who did not choose the most reasonable ballpark provided useful information about their number sense.

5. Show the students the rest of your predetermined categories and have them choose another one to explore and write about.

"OK," I told the class, "now you're going to do some work at your tables. You're going to have some choices about which problem to work on. I have three topics for you." I held up three more of the category papers:

- *Are wearing tennis shoes*
- *Had cereal for breakfast*
- *Have a younger sister*

I like to provide students with choices when possible. Offering several options rather than dictating one assignment for everyone gives students more independence and ownership. In this instance, the subject of the hypothesis didn't matter: I would be able to see the students' ideas about numbers regardless of the topic they chose.

I explained the task. "You are going to choose one of these categories and decide which ballpark it fits in. Remember, we are talking about the people in the room right now. So you need to think about what fraction of people in the room are wearing tennis shoes or had cereal this morning or have a younger sister. Spend a little time talking at your tables about each of these. Then you're going to pick one to concentrate on. Your job will

Teaching Tip

Ask students to write about their thinking; this further develops their abilities to be precise. Not only do they need to reflect on and clarify their own thinking, they need to use appropriate mathematical language to express their ideas.

Teaching Tip

Provide students with choice when possible. This gives students more independence and ownership of their learning.

be to write a convincing argument about the one you chose and explain where you think it fits. Are there any questions?"

About fifteen minutes before the lunch break, I made an announcement.

"You've had some time to talk about these three categories at your table and choose one to think about more. Now it's time for everyone to concentrate on writing your ideas. Let's take some quiet time so that everyone has a chance to get some ideas down on paper."

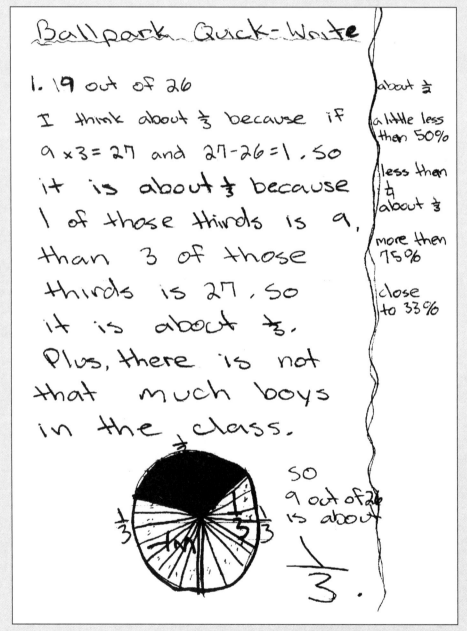

Figure FL4–1. Alicia used a pie graph to help with her estimate.

I collected the papers before the students went to lunch. Interestingly, all three choices received about equal attention. The papers showed a variety of approaches and a range of thinking. Some of the students used empirical information exclusively to explain their ballpark choice. Mary wrote, *About $\frac{3}{4}$ of the people in this class are wearing P.E. shoes because I took a peek under everybody's desk.* Calvin used his own experiences as well. He wrote, *I think about $\frac{1}{2}$ the class has a younger sister because I know 9 people and 5 of them have a younger sister. Since there's more than 9 people in the class I think it's about $\frac{1}{2}$.*

Several of the students used a broader generalization as a rationale for their decision. Although the task focused on people in the classroom, these students started with a bigger picture and then worked back to the smaller. (For some reason, this type of broad thinking occurred mostly in connection with the breakfast cereal category.) Emma was thinking nutritionally. She explained, *About $\frac{1}{3}$ of the kids in my class eat cereal for breakfast. That is because when kids get older they need more to eat, not just cereal.*

Many students worked on the tennis shoe category. Sariah included a written explanation and a picture on her paper. She imagined what the two groups (wearing tennis shoes and not wearing tennis shoes) would look like if they were on two different soccer teams. She wrote, *I think that more than $\frac{3}{4}$ of the class is wearing tennis shoes today, because if you were playing a soccer game, and you needed to have even teams. So if you took the people that were wearing tennis shoes for a team and boots, high heels, etc., in another team to make two teams it would be very uneven.*

The papers gave me an insight into the students' number sense, particularly their estimation strategies. Did they use sample data to predict? Which ballpark did they equate with which numbers? Did they compare whole numbers with fractions and decimals? How? The assessment possibilities of this activity are rich, and repeated discussions and writing assignments will definitely strengthen the students' estimating capabilities.

Formative Assessment

Looking at student work products gives insight into the thinking of individual students and also the range of ideas within the classroom.

Reflecting on the Lesson

How does this activity promote the development of number sense?

Estimation requires a feel for quantities, relative magnitude, and benchmarks. It's vital for students to have many opportunities to estimate in math class, so that they will develop these abilities and practices. This activity gives students a chance to estimate with fractions and also to discuss and write about their thinking: not only are they practicing estimation, they are being given opportunities to think about estimation in different ways. Thinking about their own methods of estimating and listening to their classmates' ideas build their estimation proficiency.

Many estimation activities involve guessing how many of something. This activity asks students to think about estimation in a different way. Presenting the students with real situations and a number of ballparks requires them to compare and order fractions. They also need to collect data or use a benchmark to help them make sense of the situation and choose the most reasonable quantification. In most instances students do some mental computation to help them choose the most reasonable estimate.

Why did you choose these particular "ballparks"?

I typically give students a mix of unit fractions (with a numerator of 1) and nonunit fractions. In the example I stuck with $\frac{1}{2}$s, $\frac{1}{4}$s, $\frac{1}{8}$s, and $\frac{1}{3}$s because once students have a solid grasp of these fractions, they are better able to negotiate the number system. They can handle unfamiliar numbers by comparing them with more familiar benchmarks. For example, some students recognized that $\frac{9}{26}$ was close to $\frac{9}{27}$ which is equivalent to $\frac{1}{3}$.

An interesting follow-up activity is to ask the students to put the ballpark choices on a number line in ascending order or to find ones that are very close to each other.

Can this activity be adapted for other grade levels?

Yes, this lesson is very flexible. Whole numbers, fractions, or decimals all work well. Third-grade teachers might want to start with whole-number benchmarks such as 25, 50, 75, 100 and find topics that fit in this range. Fourth-grade teachers working on computation with larger numbers can use benchmark numbers in hundreds and thousands to help build students' understanding of the place value system and the magnitude of numbers. Students can also generate their own ballparks or their own topics. You can devote part of a class session to brainstorming ideas of things to explore and categorize. Or you can present a subject such as *number of crayons in the room* and ask students to name a ballpark that seems reasonable to them.

Learning Stations (Games)

Section Overview

Defined by Jennifer Lempp as "activities in which students engage in meaningful mathematics and are provided with purposeful choices" (2017, 121), the games for learning stations serve as hybrids of routines and focus lessons. Number sense games help students develop and practice number sense skills in an engaging and thought-provoking context. In most instances, the games are introduced using the whole-class lesson format. As with routines, the initial investment in a whole-class introduction to the game pays off when students have the understanding and experience to play independently. The opportunities to learn a game, play it with a partner, analyze structure and strategy, and listen to classmates' thinking turn the games into rich math tasks worthy of a math class period.

GAMES

G1	One Time Only	99
G2	Oh No! 99!	110
G3	Hit the Target	122
G4	Get to Zero	133
G5	Get to 1,000 (Addition)	143
G6	Get to 1,000 (Multiplication)	153
G7	Decimal Maze	163

One Time Only

Overview

One Time Only is a great opportunity for students to develop their number sense. In this two-person game, students take turns identifying factors of successive numbers, continuing until one of them can no longer contribute a new number. The game context engages them, and the discussions they have with their partner help them think about numbers. Their experiences during the game become the basis for some substantial whole-class discussions about important topics in math. Asking students to write about the game gives them an opportunity to clarify their thinking. As the students continue to play *One Time Only*, they build on their previous experiences and further develop their number sense.

Learning Targets

GRADE 3

I can find the missing number in a multiplication or division problem.

I can find the answer to a division problem by thinking of the missing factor in a multiplication problem.

I can multiply and divide within 100.

I can find patterns in multiplication tables and explain them.

GRADE 4

I can use the commutative property of multiplication.

I can recognize a whole number as a multiple of each of its factors.

I can determine whether a whole number from 1 to 100 is a multiple of a given one-digit number.

I can notice and point out different features of a pattern.

NUMBER SENSE FEATURES

(see page xv)

Materials

None

 Time

Ten minutes to one hour (depending on version)

Teaching Directions at a Glance

1. Player 1 writes down a number greater than 1 and less than 100.
2. Player 2 writes down a factor of the first number underneath it.
3. Player 1 writes down a factor of this new number.
4. Each player, taking turns, writes down a factor of the preceding number.
5. If a player writes down a prime number (i.e., it is not divisible without a remainder by any other integers except 1 and itself), the next player adds 7 to it and writes down the resulting sum as his or her turn.
6. Play continues until no new numbers are available.

ADDITIONAL RULES

1. Once a number has been written down, it can't be used again.
2. The number 1 can't be used at all.

Teaching Directions with Classroom Insights

From a Fourth-Grade Classroom

"Today I have a game for you," I announced to Shea Carrillo's fourth-grade class. "It's called *One Time Only*. To play the game you need a partner. One of the partners begins by picking a number greater than 1 and less than 100. So you can see how it works, the whole class will be my partner for this first game. There are just a few rules, and I'll explain them while we play."

1. Player 1 writes down a number greater than 1 and less than 100.

I wrote *36* on the board.

"Now it's your turn," I said. "You need to think of a factor of 36. Can anyone tell me a number that goes evenly into 36? Another way to think about it is by skip-counting. Which numbers can you skip-count by and get to 36?"

By introducing several ways to think about factors, I hoped to explain the game more quickly. If I'd just asked for a factor of 36, students who weren't sure what a factor was or who weren't sure about the difference between a factor and a multiple might not have been able to participate. As the students use the terminology in the context of the game, they'll become more comfortable with it.

"So, what do you think?" I asked. "Can anyone tell me a factor of 36?"

2. Player 2 writes down a factor of the first number underneath it.

"How about 6?" offered Fred.

"Is 6 a factor of 36?" I asked the class.

Several students nodded or vocalized their assent. I pushed for more of a commitment in order to reinforce "Who can explain why they think 6 is a factor of 36?" I asked.

Jessie raised her hand. "Because 6 times 6 is 36," she explained.

"Also," added José, "if you count by sixes you'll say 36. Like 6, 12, 18, 24, 30, 36."

"All right," I said, "I'm convinced that 6 is a factor of 36." I wrote *6* under *36* on the board.

Mathematical Practice

Having students explain their thinking helps them with the practice, construct viable arguments and critique the reasoning of others (MP3).

3. Player 1 writes down a factor of this new number.

"Now I need to find a factor of 6 to add to the chain of numbers we're making here. I think I'll say 2." I wrote 2 under the 6.

4. Each player, taking turns, writes down a factor of the preceding number.

"Okay, now it's your turn to think of a factor of 2," I said.

"Two," said Derek. "Two times 1 is 2."

"Well, yes," I responded, "2 is a factor of itself, but one rule is that you can't use the same number twice. That's why the game is called *One Time Only*. If a number is already written down you can't use it again. Can anyone think of a factor of 2 that's not already up here?" I asked, pointing to the board.

"One," said Ali.

"Well, that brings up another rule in *One Time Only*. You can't use 1. You're correct, Ali, that 1 is a factor of 2. But in this game you're not allowed to use 1. So you can't use a number that's already up there and you can't use 1. Those are the two main rules of this game. Can you think of any other factors of 2?"

"How about 4?" asked Chrissy.

"How do you know 4 is a factor of 2?" I inquired.

"Because 2 times 2 equals 4," Chrissy explained. Chrissy had confused factors and multiples. I was glad she had made the multiplication connection, but I needed to prompt her a bit to get her back on track.

"I know that 2 is a factor of 4, because I can count to 4 by twos," I said to her. "But it doesn't work the other way around. Four isn't a factor of 2, because you can't count to 2 by fours."

"Oh, yeah," Chrissy replied.

"Does anyone know what we call 4 in this situation?" I asked the class.

"A multiple!" exclaimed Neal. "If you can times a number to get the number it's a multiple. Like 36 is a multiple of 6 because 6 times 6 is 36."

"Alright," I continued, "so are there any factors of 2 besides 2 and 1?"

I kept posing the question to get students really to think about the number 2 and its relationship to other numbers. This sort of thinking builds number sense. Also, I wanted the students to convince themselves that 2 only has two factors. "Take a minute and talk at your tables," I suggested. "See if you can think of any other factors of 2." I let the students talk briefly and then I called them back to attention.

"Did any tables find any other factors of 2?" I asked. The class consensus was no.

5. If a player writes down a prime number, the next player adds 7 to it and writes down the resulting sum as his or her turn.

"Well," I told the class, "you're right. There are only two factors of the number 2, 2 and 1. Does anyone know what you call a number that only has itself and 1 as factors?"

"Prime?" Greg ventured in a barely audible tone.

"Prime!" several students announced with authority after hearing Greg.

"Yes, those are prime numbers." I wrote *prime* next to 2 on the board. "In *One Time Only* when you hit a prime number you add 7 to it. So what's 2 plus 7?"

"Nine," several students responded. I wrote *9* on the board under the 2.

"Okay, now it's my turn, and I need to think of a factor of 9. I'll say 3," I said, as I added *3* to the list on the board. "Now you need to find a factor of 3."

"It's prime," announced Natalie with authority.

"Are you sure about that?" I asked the rest of the class.

"Yes, it is," agreed Jasper, "because 3 times 1 is 3 and that's it."

"If it's prime, what happens?" I asked.

"Add 7," José reminded us. "So it's 10." I wrote *prime* next to the 3 and put a *10* below it.

"How many times can you use plus 7?" Jessie asked.

"There's no limit," I explained. "Anytime you're playing and a prime number comes up, you just add 7 to it. It's my turn again, and I need to put a factor of 10 that's not already listed. I'll say 5."

"Oh no," exclaimed Alejandro, "another prime number for us."

I raised my eyebrows in feigned surprise as I looked at the numbers on the board. "Wow, it *is* a prime number." I agreed. "You keep getting prime numbers on your turn. I wonder if that always happens in this game. Maybe there's some kind of pattern."

While I knew that this particular pattern didn't always happen, I took the opportunity to spark a little curiosity. I hoped that in subsequent games students would pay more attention to the occurrence of patterns in general as they played. Looking for patterns is a powerful way to build number sense, particularly when students have opportunities to think about the patterns and their relationships to numbers and operations. It is also a key aspect of the math practice which asks students to look for and make use of structure. The more opportunities we can find to build

Teaching Tip

Add 7 to the prime numbers to extend the game; otherwise it would end when the first prime number occurs.

Mathematical Practice

Highlighting emerging patterns and hypotheses develops students' inclination to expect and seek out patterns and structure in their work with numbers and operations (MP7).

pattern seeking, pattern recognition, and pattern description into our math lessons, the more this practice becomes second nature.

I referred to the string of numbers on the board, which now looked like this:

36

6

2 prime

9

3 prime

10

5 prime

"Okay," I said to the class, "it's your turn and since the number 5 is prime, what do you need to do?"

"Add 7," Jasper replied.

"Right," I agreed, "so now it's 12." I wrote *12* on the board. "Hmm," I said, "I need to find a factor of 12 that's not already up here." I paused for a few seconds to give students a chance to review the numbers and think about factors of 12. I also wanted the students to see that math involves taking time to think.

"I know," I brightened, "I'll say 4." I wrote *4* on the board beneath the *12*. "Now you need to find a factor of 4 that's not already up here. Talk at your tables for a minute or two and see what you can come up with."

6. Play continues until no new numbers are available.

"We're stuck," Ali soon announced.

"What do you mean?" I asked.

"Well," Fred explained, "we're not allowed to use 1. Four and 2 are used already. There are no other factors, so we can't go."

"Does everyone agree with Fred and Ali?" I asked, looking around.

The nods and *yeahs* were unanimous.

"Then I guess the game is over," I said. "This time I won, because I was the last player to add a number to the list. You want to get your partner stuck so she or he is unable to add a number to the string. But winning isn't really the important part of this game. You're going to play a bunch of times, and sometimes you'll win and sometimes you'll lose. The important part of the game is the mathematical thinking that you do."

"Raise your hand if you understand *One Time Only* and you're ready to play with your partner," I instructed the class. The students were ready. "I want you to know the plan for the rest of this math class," I told them. "You definitely need to have some more time to play *One Time Only* so you get familiar with it. Before the end of class we're going to get back together and have a discussion about your experiences playing the game. I'm interested in hearing about any strategies you used to help you win the game. Maybe you'll notice some patterns and predictable things that kept happening during the games. We'll also talk about any tricky numbers that came up while you were playing."

"What do you mean?" asked Katrine.

"By tricky numbers I mean numbers that aren't so easy to find factors for," I elaborated. "Some numbers are pretty easy because they come up a lot on the multiplication table. But there are other numbers between 1 and 100 that you don't use very much when you're doing basic multiplication facts—47, for example. It will be interesting to hear which numbers you ran into that were kind of tricky and how you found factors for them."

As the students began to play, I visited the tables. Most of the children had paired up and begun playing without much prompting. The whole-class game we played had three prime numbers in it, but Katrine and Shante excitedly reported they had just finished a game in which four prime numbers appeared. I announced this milestone to the class at large. Conversations erupted throughout the classroom about prime numbers and how many times they had been encountered in different games.

Ana and Ronald were stuck on 97. Ana got a calculator from the shelf.

"Divide it by something," Ronald told her.

Ana punched in 97 divided by 3 and got 32.33. "What does that mean?" she asked Ronald.

"It's not right—you can't divide it by 3. There's a decimal. That means it didn't divide evenly." Ronald explained. "Try 7."

Ana found that 97 divided by seven was 18.85. "Nope," she told Ronald. The pair continued to guess and check by dividing 97 by 6 and 4. "I think it's prime," was Ana's appraisal.

"It is," agreed Ronald.

"Are you sure?" I asked. "What about dividing 97 by 5? You didn't try that."

Teaching Tip
Let students know in advance about the topic for a summary discussion; this helps them prepare and be ready to share their discoveries.

Teaching Tip

Perhaps one of the best question in math class is "Are you sure about that?" It prompts students to reflect on the mathematical reasonableness of their answer and construct arguments to defend it.

Teaching Tip

Call the class back together for a whole-class discussion; encourage students to share their strategies.

"Well," Ana explained, "5 will definitely have a decimal, because 97 doesn't end with 0 or 5. When you count by fives the numbers always end with 0 or 5."

Impromptu discussions like this are an excellent opportunity to help students build their number sense. Ronald and Ana were thinking about numbers, their relationships and patterns, and the implications of a decimal. As I continued to circulate, I tried to help the children focus on the numbers.

Fred and José were eager to start with a number greater than 100. I asked them to stick with numbers less than 100 for the moment, but I agreed that using larger numbers would be an interesting investigation for the future. Having options and challenges is an important aspect of any well-differentiated whole-class lesson.

I called the class back together with about fifteen minutes left in the period.

"Let's talk about your strategies," I suggested. "For example, when Elliot was playing, he told me he thought it was a good idea to start with an odd number. This was his strategy to help him win the game. Did anyone have any other strategies that seemed useful?"

Natalie's hand shot up. "I concentrated on 6."

"What do you mean?" I asked.

"Well," she explained, "I tried to end with the number 6. But first I tried to get rid of all the factors of 6, like 3 and 2, so that way when I wrote down 6, I knew my partner couldn't do anything. Also, if you get rid of the factors of 6, you're also getting rid of the factors of 9 and 4." I was pleased to see that Natalie had indeed looked for and found structure in the game.

"It worked," said Natalie's partner, Juanita.

"What other strategies did you use?" I asked the class.

"Try to end with an even number," advised Jessie.

"Start with an odd number," Katrine added.

As the students shared their strategies, I recorded each one on the board. Alejandro had an idea to add to the list.

"My strategy was to get rid of 3 and then give my partner 9."

Natalie had more to say. "This isn't a winning strategy," she qualified as she began, "it's a winning pattern. In our games we had this pattern, 2, 9, 3, 10, 5, 12, 6."

"Hey," interjected José, "we had the same pattern!"

"We had almost the same pattern," Shante piped in. "But ours is 2, 9, 3, 10, 5, 12, 4."

"This is very interesting," I confirmed. "As you play more I wonder if you'll find more patterns. I'm also wondering if seeing these patterns can help you predict who's going to win."

These were not questions that could be answered during the current whole-class summary discussion, but I was confident they would be explored more intently over time. The students needed more experience playing the game and thinking about the implications of these patterns— why they emerged and how they might be manipulated. Often a whole-class discussion is not a neat and tidy wrap-up but rather an occasion to raise questions that lead into deeper mathematical territory. Leaving some questions unanswered also develops students' skills in making sense of problems and persevering in solving them. Interesting math questions are not always answered in one sitting.

"There's one other thing I want to ask about," I continued. "Did anyone encounter a tricky number? What were some of the numbers that were hard to find factors for?" I gathered a bunch of responses quickly and wrote all the numbers on the board under the heading "Tricky Numbers." The list included *75, 26, 47, 89, 11, 68, 17, 44, 59, 99,* and *62.*

Having students write about their thinking and strategies helps them become proficient with precise language. While the word *precision* in the context of math readily brings precise calculations to mind, this mathematical practice also refers to being precise with language and using proper mathematical language and terminology.

"I am really interested in knowing more about your thinking," I told the class. "Therefore, I have a homework assignment for you. There are just two questions for you to answer, but you need to try to explain your answers fully so I'll understand how you thought about these questions. The first question asks you to think about the number 68 and different ways you can find the factors of 68. The other job is for you to teach *One Time Only* to someone at home and write about the strategies you used to try to win."

I chose the number 68 because I knew it had factors but it wasn't a number that occurs in the context of typical multiplication tables. The students would need to think about quantities and their relationships rather than just spit back some memorized numbers.

The students had many methods for finding factors. Several of the students outlined specific systematic approaches to the task. Others described a trial-and-error process. Rolanda's response (on the following page) to the first question showed she was able to take 68 apart in several helpful ways:

Mathematical Practice

Keep in mind that a whole-class discussion might not be a neat and tidy wrap-up; this is OK. Leaving some questions unanswered develops students' skills in making sense of problems and persevering in solving them (MP1).

Mathematical Practice

Having students write about their thinking and strategies helps them become proficient with precise language (MP6).

Formative Assessment

A homework assignment related to the day's activity gives students an opportunity to reflect on what they learned (self-assessment).

1. *If it is an even number then you would be able to split it in half.*

2. *You could times 4 by 10 and get 40. Then 4 times 5 is 20, so that's 60 and then we all already know that 2 fours is 8, so that's 68.*

3. *Another easy way is to just divide 68 by another number, and if you get a whole number, then you have a factor: 2, 4, 34, 17, 68. I know that all these are the factors of 68 because I used my methods on all of the numbers smaller than 34 because any number larger than 34 is not a factor. They are not factors because any number larger than 34 would go over 68 if you times it by 2.*

The descriptions of strategies were enlightening as well. I was able to see the different ways students thought about the game and the different ways they expressed themselves in writing. Shannon gave a detailed plan with an explanation for each step:

If you go first pick a prime number. Then the opponent will have to add seven. It will become an even number. Your turn, divide the number by half of the number. You will get 2. The number is prime again. The opponent will now have to add 7. You will get 9 (7 + 2 = 9). Your turn again. Divide 9 by 3. You will get 3. Another prime number. Again the opponent will have to add 7. You get 10 (3 + 7 = 10). Divide it by 2 and you get 5. Once again a prime number. The opponent adds 7 (5 + 7 = 12). You get 12. Now divide 12 by 2 and you get 6. For once it is not a prime number for the opponent. The factors for this number that are allowed are 2 and 3. But wait a minute. We already used 2, and 3. So the person who started this game won.

Reflecting on the Lesson

When would you play this game with third graders?

I wouldn't introduce this game to third graders at the beginning of the year. It's important that they have a good foundation in multiplication first. If the students are unfamiliar with the concept of multiplication and the relationship of factors to products, then *One Time Only* could be frustrating or turn into a guessing game.

However, once third graders have a solid grasp of multiplication and division, *One Time Only* is an excellent way for them to reinforce the basic facts while exploring ways to think about factors of different numbers. With third graders, spend extra time talking about ways to find factors before sending them off to play on their own. Providing them with manipulatives and reminding them of the array model of multiplication could be helpful scaffolds. That way they have some tools available if they get stuck.

What do students get from this game if they already know their multiplication facts?

In addition to practicing multiplication facts, students who play *One Time Only* are exploring the number system and learning about its underlying structure. Thus, they are getting rich experience with mathematical practices. As students play the game, they begin to develop strategies. In order to devise a winning strategy, students must consider the relationships of numbers. For example, Brittany noticed that the factors of the number 6 are also factors of 4 and 9. Seeing these connections between numbers helps build number sense.

Also, as students examine their records of the games they've played, patterns emerge. These patterns are another opportunity to uncover structure and develop number sense. Teachers can ask their students to think about why certain patterns recur, what causes these numbers to arise in a specific order, and how the patterns can help predict the outcome of a game. The game provides many occasions for writing, as well. It might be interesting to add a number other than seven to the primes and see how that affects the outcome. It would also be fruitful to explore the effects of adding 7 (or other numbers) to a prime number. This exploration leads to discoveries about prime, composite, odd, and even numbers.

Amanda and Ronald used a calculator. Is this appropriate?

Amanda and Ronald were trying to find factors of 97. When their knowledge of basic multiplication facts failed them, they used a calculator. I felt fine about it. A calculator is a tool to be used in service of a problem. They used an appropriate tool to strategically answer a question in service of a larger problem. Amanda and Ronald had a problem and knew that dividing 97 by different numbers would tell them something about its factors. When Amanda and Ronald couldn't think of any factors for 97, I could have jumped in and told them that 97 was prime. Instead, I gave them the freedom to delve further. It turned out to be a double learning opportunity. In addition to convincing themselves that 97 was prime, they also confronted decimals and had to make sense of their meaning.

GAME

Oh No! 99!

Overview

While older elementary students are typically engaged with larger whole numbers, many still need and benefit from practice with mental addition and subtraction of smaller numbers. In *Oh No! 99!*, a two-person card game, players attempt to force their partner to be the one to push their jointly accumulating score above 99. The game provides practice with adding and subtracting while also giving students the chance to think strategically.

NUMBER SENSE FEATURES

(see page xv)

Learning Targets

GRADE 3

I can use mental math to figure out the answer.

I can quickly and easily add and subtract numbers within 1000.

GRADE 4

I can understand and explain the value of digits in numbers.

Materials

A deck of playing cards (jokers removed) for each pair of students
Hundreds charts (as needed)

🕐 Time

Forty-five minutes to one hour

Teaching Directions at a Glance

1. One player shuffles the cards and deals four cards to each player. The undealt cards remain in a stack, face down.

2. Players take turns playing one card at a time, adding or subtracting the value of their card to or from their jointly accumulating score. Each time a player plays a card, he or she must replace it with the top card from the face-down stack.

3. Play continues until one player forces his or her partner to go over the score of 99.

CARD VALUES AND OPERATIONS:

1. Aces: add 1
2. Jacks: subtract 10
3. Queens: wild cards that can represent any other card in the deck
4. Kings: add 0
5. All others (2–10): add their face value

Teaching Directions with Classroom Insights

From a Third-Grade Classroom

To introduce *Oh No! 99!* to an enthusiastic group of third graders, I'd planned to go through a sample game with the whole class and then send them off to play with partners. I began by asking the class to join me in a circle on the rug, so everyone would have a good view of the deck of cards. After some adjusting of desks and bodies, we were ready to begin.

"I brought a lot of decks of cards here today because you're going to learn a card game," I announced. I had the students talk to a partner about what they know about playing cards. This little discussion tapped in to their prior knowledge and surfaced some useful vocabulary.

I walked to the board and wrote:

Oh No! 99!

"This game is called *Oh No! 99!* As I explain the game you might start to get some ideas about how it got its name. Now before I show you how to play, there are a few important things you need to know about the cards." I wrote *A, J, Q, K* on the board.

"In *Oh No! 99!* an ace means add one point," I explained. I wrote *add 1* next to the *A*. "A jack means subtract 10. A queen is wild. That means you can assign the queen the value of any other card in the deck."

"It can be any number?!" asked Jeannette, wide-eyed.

"Can it be 1,000?" asked Kenneth.

"Well, it can't be 1,000 because there's no other card in the deck worth 1,000. It's wild, but it's not that wild," I told the class.

I proceeded with my explanation. "A king means you don't add or subtract anything. For the rest of the cards in the deck, add whatever their number is: an 8 means add 8, a 3 means add 3. I'm going to leave this information up on the board, because you might need it when you play with your partner. The object of the game is to make your partner go over 99. So, I want to force you to put down a card that makes the total 100 or more, and you want to try to get me to do the same. Can you guess why this game is called *Oh No! 99!*?" I asked.

"Because if there's 99 already, you're in trouble," offered Evelyn.

"You got it," I agreed. "Before you go back to your tables, though, we're going to play part of one game together, so you can see what the game is like, and I can answer any questions. Since this is a game for partners, I'll play with the whole class as my partner. We're going to take turns adding cards to the discard pile."

Teaching Tip

Give students a little time to talk about new material; this cuts down on off-task behavior later in the lesson.

1. One player shuffles the cards and deals four cards to each player. The undealt cards remain in a stack, face down.

I dealt four cards to myself and four cards to the class. I dealt all the cards face up, although I explained that when students play by themselves, they keep their cards hidden so their partner won't know what they have.

2. Players take turns playing one card at a time. Each time a player plays a card, he or she must replace it with the top card from the face-down stack.

"Okay, I'll go first, and I'll put down this 7. And since I put down a card, I need to pick another card from the top of the deck. I always want to have four cards in my hand. Now it's your turn. Who would like to choose a card for the class?" Many hands shot up. I called on Gina.

"I'll use the 9," she announced.

I put the 9 on top of the 7.

"So now what's the total for the pile?" I asked the class.

"Sixteen!" they responded in unison.

"Good," I said, "Seven plus 9 is 16. Whenever you put a card down on the pile you have to say the whole equation to tell what the new total is. Partners need to check each other and pay attention to make sure both of you know the total."

We continued playing. I called on various students to make a choice for the class, and had the whole class say the equation and tell me the new total after each card was added. When the total of the pile was 88, I asked Jenny to choose a card for the class to play. Many students tried to influence her.

"Don't use the ace yet," advised Miguel.

"Put down the 9," suggested Annabel.

Realizing that the advice was being motivated by strategic thinking, I stopped the game to point this out.

"I'm noticing that many of you have ideas about which card to play next. Would anyone like to explain your thinking?" I called on Kate.

"I think we should save the ace."

"Why is that?" I asked.

"Because with an ace you only have to add 1, and that's a low number. If the cards get up to a high number like 97 or 98, we can use the ace to make you go over 99." Many students nodded in agreement.

"Okay," I continued, "Kate thinks you should hold on to your ace and save it for later. Does anyone have an idea about which card you might want to play next?"

"Use the 9," said Ana, "because then the total will be 97 and that's close to 99. If you don't have a low card or a jack, queen, or king, we could win."

3. Play continues until one player forces his or her partner to go over the score of 99.

The class played the 9, and my next play put the score over 99. I then sent the students off to their tables to play the game in pairs.

"Remember," I told them, "this game is important for two reasons. First, it gives you a lot of practice adding in your head. Second, when the total gets close to 99, you have to do a lot of thinking to plan a good strategy."

The students returned to their seats, and I circulated around the room.

Mathematical Practice

Asking students to explain their thinking helps them develop their abilities to construct viable arguments and critique the reasoning of others (MP3).

Mathematical Practice

Attending to precision (MP6) requires students to be clear and effective in both their language and their computation. Conversations like these support the practice and help students build the requisite habits of mind.

Teaching Tip

Observe the students. Is everyone understanding the game? Playing with a partner? Are students calculating mentally?

Teaching Tip

Use visual and geometric models of our number system (such as a hundreds chart) to help students internalize patterns and relationships.

At first, I just wanted to make sure everyone understood the game and was playing with a partner. Then I spent some time observing individual games. Several students were quite animated and couldn't resist showing their cards to friends nearby. Miguel, for one, was proudly flashing his picture cards to anyone in his vicinity. I issued a few gentle reminders for students to stay in their seats and focus on the game.

I noticed that while many students were quickly and easily calculating the totals mentally, others were more hesitant; some were even using their fingers. I was surprised to see Elliot use his fingers to add 10 to 43. Adding 10 should come automatically to most third graders. Elliot needed more experiences with "the big picture" so he wouldn't just fall back on finger counting or a procedure. I offered Elliot a hundreds chart as a tool to notice how to add tens. Just jumping down one square from any number on the hundreds chart adds 10. If Elliot could internalize this geometric model of the number system he could more flexibly and fluently add and subtract within 100.

Even though everyone was still very involved in playing the game, I called the class back together after about fifteen minutes. I wanted to see what kinds of strategies they were using at this point, and I wanted them to have the opportunity to hear some of their classmates' thinking about the game so far. I presented a hypothetical situation.

"Imagine," I said, "that you're playing *Oh No! 99!* and the total is up to 87. Your four cards are a 6, a queen, an ace, and a king. Which card would you play next? As you think about this, pay attention to why you're choosing a particular card. I'm going to give you ten minutes of quiet writing time so you can tell me your ideas on this question. Make sure you put your name and date on the paper. Are there any questions?"

"You just want us to tell you which card we would use?" asked Jon.

"That's part of it," I answered, "but I also want to know why you would choose that card instead of any of the others. You might even want to tell me which card you definitely wouldn't want to use and why."

Most students chose either the 6 or the queen as their next card on the pile. Traci wrote, *I would put the 6 down because you should get rid of your high cards and save your low cards as you get in the high 80s and 90s. A, Q, K, are not high cards because a Q is a wild card and you can use it as a K. A K is a 0. An A is a 1.* Neal made a convincing argument for the queen (see Figure G2–1).

After reading through the class papers (additional examples are shown in Figures G2–2 and G2–3), I realized the question didn't dig deeply enough into the methods the students used for adding the numbers.

Formative Assessment

Getting a class set of responses to the same question gives us a view of the range of responses and also shows how individual students are thinking and communicating.

OH NO 99

What would you do? I would put down the queen as a ten so the total would be 97. 97 is a high number and if your partner has only numbers higher than four you win. If they have numbers less than four or a king, queen, jack, or ace and they lay down a queen as a two or a two, you can put down the king

87 Total

Your Cards: [6] [Q] [K] [A]

Figure G2–1. Neal's hypothetical strategy.

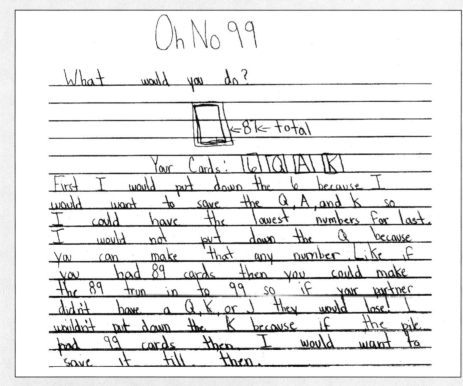

Oh No 99

What would you do?

←8k← total

Your Cards: [6][Q][A][K]

First I would put down the 6 because I would want to save the Q, A, and K so I could have the lowest numbers for last. I would not put down the Q because you can make that any number. Like if you had 89 cards then you could make the 89 turn in to 99 so if your partner didn't have a Q, K, or J they would lose! I wouldn't put down the K because if the pile had 99 cards then I would want to save it till then.

Figure G2–2. Another student's strategy.

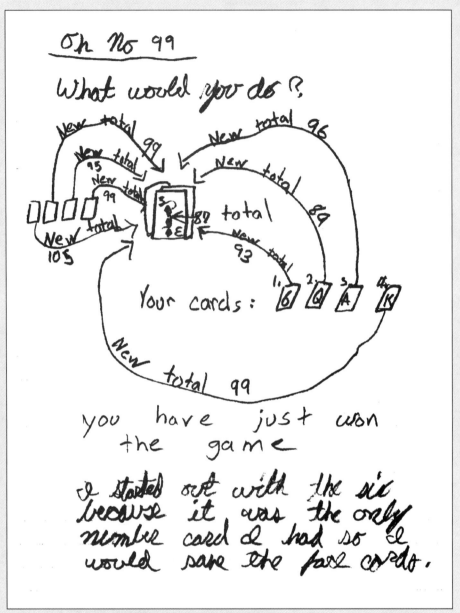

Oh No 99

What would you do?

New total 99
New total 96
New total 95
New total 89
New total 99
99 total
New total 105
New total 93
New total 89

Your cards: 1. 6 2. Q 3. A 4. K

New total 99

you have just won the game

I started out with the six because it was the only number card I had so I would save the face cards.

Figure G2–3. Still another student's strategy.

I got a general feel for their thinking about the cards, but the prompt I used focused more on strategy. I wanted to ask a question about the game that encouraged the students to tell me more explicitly how they were combining numbers. Did they use what they knew about place value to help? Were they merely counting on? Did they have more than one way to add numbers? I decided I would try to focus on these questions next time.

When I returned to the class a few days later, student partners were playing the game with gusto. I gave them a few minutes to finish, and then

I called for their attention. I began by talking about the responses they'd written during my previous visit. I told them I was impressed with both the range of their ideas and the way they were able to write about their thinking. I began the discussion with a number talk in context.

"The last time I visited I gave you a situation to think about. Does anyone remember what the total was for the question I asked?"

"It was 87," replied Charles. I wrote 87 on the board.

"And the cards we had were 6, ace, queen, king," added Lucy.

"Okay," I continued, "quite a few of you wrote that you would use the 6 and add it to the pile." I wrote + 6 next to the 87 on the board. "So what would the total be now?"

"Ninety-three!"

I wrote = 93 on the board to complete the equation.

"Now comes the interesting part," I told the class. "You all know that 87 plus 6 equals 93, so that's not really a problem. The challenging part is to think about how you solved the problem in your head without paper and pencil. Does anyone want to try to describe how you solved it mentally?"

Traci volunteered first. "I took 2 away from the 87 to make it 85. Then I took one away from the 6 to make it 5. I know 85 plus 5 equals 90. Then I just added back the 2 and the 1 and I got 93." As Traci talked through her thinking, I recorded the corresponding equations on the board:

$$87 - 2 = 85$$

$$6 - 1 = 5$$

$$85 + 5 = 90$$

$$2 + 1 = 3$$

$$90 + 3 = 93$$

Often children don't realize that the words and ideas come first and that the equation is a shorter way to express those same ideas. Many children only see equations printed in textbooks and worksheets and don't connect them with any real-world context.

As a teacher, part of my job is to help students connect equations and symbols with meaningful contexts. Number talks accomplish this goal. I asked the students to volunteer other approaches to solving 87 + 6. While the problem itself is not particularly challenging for third graders, explaining it verbally is. I wanted the students to focus on their own thinking without getting bogged down by the computation.

Teaching Tip

Continue the activity through the use of a number talk (see Routine 2).

Teaching Tip

Encourage students to see the connections between their thinking and the mathematical symbols that represent it.

Ronald shared his strategy next. "Well, I know 3 plus 3 is 6. So I took one of the threes and added it to 87. That made 90. Then I added the other 3 to the 90 and got 93." I recorded Ronald's thinking symbolically:

$$6 = 3 + 3$$

$$87 + 3 = 90$$

$$90 + 3 = 93$$

Then I recorded Jenny's, Josue's, and Evelyn's thinking on the board as well. The class got to hear five different students talk about their thinking and were able to see there is more than one way to solve a problem. This idea is a big leap for students who have been accustomed to an algorithmic approach to mathematics. The more number talks they have to expand their computational horizons, the better.

Next I wanted each student in the class to have an opportunity to construct viable arguments and critique the reasoning of others. I chose a similar context.

"Imagine you're playing a game of *Oh No! 99!* and the total score so far is 74." I wrote *74* on the board. "Your partner adds an 8 to the pile." I wrote *+ 8* next to the 74. "What's the new total?"

"Eighty-two," the class responded in unison.

"Okay," I told them, "now for the challenging part. I'm going to give each of you a piece of paper and on that paper you need to try to explain different ways you can add 74 plus 8."

Teaching Tip

Provide a writing assignment to further encourage students to think about the use of multiple strategies.

I referred to our previous discussion. "We saw five different ways people solved the problem 87 plus 6. Your job is to try to think of a lot of different ways to solve 74 plus 8. Maybe if you look at the ideas that Traci, Ronald, Jenny, Josue, and Evelyn had it might help you. Probably there are even more ways to solve these kinds of problems."

"How many ways are we supposed to get?" asked Cornelius.

"I don't have a specific number in mind," I told him. "Just try to stretch your brain to think of a lot of different ideas."

I handed out paper and the students began their work. I circulated and observed. There was quite a range of approaches. I noticed that several of the students were writing prolifically and not using symbols. While I was pleased that they were comfortable incorporating writing into their math work, I decided to steer them in a different direction.

"I'm sorry to interrupt you in the middle of your work," I told the class, "but I'm noticing that some of you are doing a lot of writing. It's great to write and use words to explain your thinking, but you can also use shortcuts and write some equations on your paper." I referred once again to the board with the five students' work. "You see I used equations to show how Traci took 2 from 87 to get 85. On your papers there can be a combination of words and equations to show how you solved the problem. Don't feel that you need to use only words for this assignment."

The students resumed working. I sat down briefly at different tables, observing students work and occasionally asking them a question about what they are writing. Several students described counting on as their strategy of choice. Romolo wrote, *This is how I did the adding. I add the number 8 into the pile. I just counted 74 and then I said 75, 76, 77, 78, 79, 80, 81, 82.* He had not recorded any other strategies. Kate had two ideas: *1. I knew the answer by counting on my fingers. 2. I added [standard algorithm with carrying].*

Shannon was clearly comfortable breaking the numbers apart and putting them back together. He had listed eleven different ways to do so (see Figure G2–4).

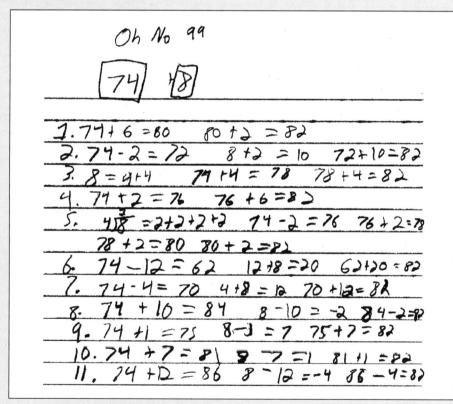

Figure G2–4.   Shannon discovered eleven ways to add 74 and 8.

Jenny used an impressive combination of words and equations to explain her thinking (see Figure G2–5).

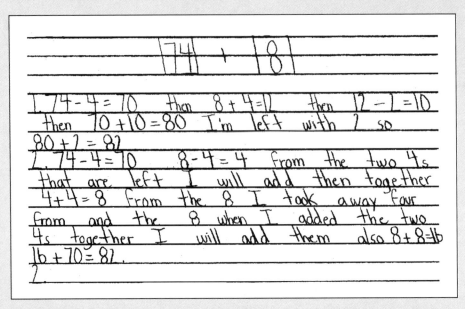

74 + 8

1. 74 - 4 = 70   then 8 + 4 = 12   then 12 - 1 = 10
then 70 + 10 = 80   I'm left with 2   so
80 + 2 = 82
2. 74 - 4 = 70   8 - 4 = 4   from the two 4s
that are left   I will add then together
4 + 4 = 8   from the 8 I took away four
from and the 8 when I added the two
4s together I will add them also 8 + 8 = 16
16 + 70 = 82.

Figure G2–5. Jenny's ways to add 74 and 8.

Jon used 10 as a friendly number: he added 10 to 84 and then took two away.

I got the feeling some of the students were humoring me. They listed different ways to make 82, but didn't seem concerned with using the original numbers or context. It was still good to see they were thinking flexibly and using all operations to decompose and recombine the addends. The great thing about *Oh No! 99!* is that students are happy to play more than once. There would be many opportunities for practice, follow-up, and refinement.

Reflecting on the Lesson

What is the purpose of this activity?

Oh No! 99! provides a context in which students can practice mental computation. During the game, the students are repeatedly adding and occasionally subtracting numbers between 1 and 99. This certainly gives them the practice they need to become computationally proficient and efficient.

Additionally, the game gives students an opportunity to use strategic thinking while they are playing. Students need to consider the value of the cards in their hand, hypothesize about what cards their partners might have, and make decisions based on their ideas.

Finally, the game provides a context in which students can do some written work. Their writing gives insight into how they communicate mathematically and how they think about breaking numbers apart and putting numbers together. It's important for students to have many opportunities to practice these skills.

Is this game too easy for third graders? What about using it in fourth or fifth grade?

On the surface this is an easy game, but there is a lot of math embedded in it. Students playing *Oh No! 99!* are practicing mental computation, using strategic thinking, and communicating their thinking. One test of the game is how engaged the students are. If it were truly too easy, the students would lose interest rather quickly. These students continued to choose to play *Oh No! 99!* during their free time throughout the year, indicating it offered an appropriate challenge.

Teachers can also put the lesson emphasis on communicating, strategic thinking, and mental computation. In any event, providing engaging computation practice for upper-elementary students helps build and maintain number sense.

Hit the Target

Overview

While it is important for students to learn to multiply accurately, it is equally important for them to estimate answers to multiplication problems. When students can estimate, they are better able to judge whether answers in whatever way obtained are reasonable. In *Hit the Target*, students work in pairs. They figure mentally or use a calculator to multiply numbers together to produce a product that falls within a predetermined range. The goal is to hit the target in as few steps as possible. (In a more advanced version of this activity, students often need to multiply using decimals—see Extension 1.)

Learning Targets

**NUMBER SENSE
FEATURES**

(see page xv)

GRADE 3

I can multiply any one-digit whole number by a multiple of 10 (6×90, 4×30).

I can determine how reasonable my answers are using mental computation, estimation, and rounding.

GRADE 4

I can multiply a whole number up to four digits by a one-digit number.

I can multiply two two-digit numbers.

I can determine how reasonable my answers are using mental computation, estimation, and rounding.

GRADE 5

I can easily multiply larger whole numbers.

I can determine how reasonable my answers are using mental computation, estimation, and rounding.

I can add, subtract, multiply, and divide decimals to hundredths using what I have learned about place value.

Materials

A calculator for each pair of students

 Time

Thirty to forty-five minutes

Teaching Directions at a Glance

1. Players choose or are given a target range (800–850, for example), in keeping with the kinds of numbers they are comfortable with.

2. Player 1 chooses a number between 1 and 100 (50, for example).

3. Player 2 chooses another number to multiply the first number by, either mentally or with a calculator (50×10, for example), and Player 1 verifies and records the result.

4. If the product doesn't hit the target range, Player 2 goes back to the original number and multiplies it by another number (again, either mentally or with a calculator), and Player 1 verifies and records the result.

5. Players repeat Step 4 until the product falls within the target range.

6. Players repeat the game, this time alternating roles.

Sample Game Scenario

Target Range:	800–850
Starting Number:	50
$50 \times 10 = 500$	The number is too low.
$50 \times 20 = 1,000$	The number is too high.
$50 \times 15 = 750$	The number is closer but still too low.
$50 \times 17 = 850$	The number is within the target range.

Extensions

1. If the product doesn't fall within the target range, students use the product (not the original number) as their new starting number and determine what number to multiply it by to hit the target range. This version of the game often involves multiplying by decimals to get to the target. Before playing it, students should have spent some time exploring decimal numbers with calculators, seeing what happens when they multiply a number by another number that is less than one, what happens when they multiply a number by 1.5, etc.

2. Play *Hit the Target* using addition and subtraction rather than multiplication.

Teaching Directions with Classroom Insights

From a Fifth-Grade Classroom

"I have a new game I want to teach you called *Hit the Target*," I said, as I began the activity with Pam Long's class. "In this game, you'll use mental math and a calculator to multiply numbers."

1. Players choose or are given a target range.

"The game is called *Hit the Target* because the goal is to hit a target range—800 to 850, for example—by multiplying a starting number by some other numbers. The idea is to do this in as few multiplications as possible." I placed a piece of paper and a calculator where everyone could see it. I wrote *Hit the Target* at the top of the paper, and then wrote *Target Range: 800–850*.

"Two people play," I continued. "Player 1 picks the starting number and Player 2 mentally figures out what to multiply by to get into the target range. When you play, you and your partner take turns being Player 1 and Player 2. Would someone like to play *Hit the Target* with me?" Lots of hands wiggled in the air. "You be Player 1, Mindy, and I'll be Player 2," I said. "To play, we both have a few jobs." I pointed to the directions, which I'd written on the board beforehand.

2. Player 1 chooses a number between 1 and 100.

After reading the directions out loud, I said, "Mindy, give me a number between 1 and 100 to start with."

"How about 2," she suggested, writing 2 on the piece of paper so that everyone in the class could see.

3. Player 2 chooses another number to multiply the first number by, and Player 1 verifies and records the result.

"Now I have to think of a number, any number, so that 2 times the number will give an answer between 800 and 850," I said to the class. "Raise your hand if you have a suggestion for me."

"Try multiplying by 425," said Gordon.

"Why did you suggest 425?" I asked. Asking students to explain their thinking helps them justify their ideas and communicate them to others.

"Because if you double 425, it's 850. It's easy," he answered. Mindy wrote down $425 \times 2 = 850$.

"Gordon already checked the answer for me using mental math," she said. "You hit the target in one move!" Mindy returned to her seat.

4. If the product doesn't hit the target range, Player 2 goes back to the original number and multiplies it by another number, and Player 1 verifies and records the result.

"Thanks, Mindy," I said. "If Mindy and I were to continue with the game, we'd switch and I would be Player 1 and she would be Player 2. But this time, I'll be Player 1 and the class will be Player 2. I'm going to choose 12 to start." I wrote down the number *12*. "Now you have to think of a number to multiply by 12 to hit the target range (800–850)," I reminded them. "Any suggestions?" The room was quiet. After several seconds, I asked a question to stimulate their thinking. "How about multiplying 12 by 5? Do you agree or disagree? I asked.

Asking students if they agree or disagree with another student's (or teacher's) thinking encourages everyone to evaluate what was said and provide further reasoning for its accuracy or inaccuracy.

"I disagree because that's only 60," said Anita. "You have to try a much bigger number."

"Well, how about 100 then?" I inquired. Students were shaking their heads in disagreement. "Why shouldn't you do that?" I probed.

"Because 12 times 100 is 1,200, and that's way over the target range," Katie said. "Let's try 12 times 50."

"Okay," I said to the class. "Multiply 12 times 50 mentally." I knew this would be challenging for some students, so I gave the class time to think about the problem. After more than half of the students had raised their hands, I called on Xavier.

"It's 600, because 12 times 5 is 60, and you have to do ten times more than that, and that would be 600," he reasoned. Knowing how to multiply by 10 and multiples of 10 is an important strategy when calculating mentally—and a signal of having number sense.

Jenny did it another way. "Ten times 50 equals 500, and two more 50s make 600."

"I thought about it like this," Mindy said "Twelve times 10 is 120. Then you need five 120s. One hundred and twenty and 120 is 240; that's two. Two hundred and forty and 240 is 480; that's four. So 480 and another 120 is 600, because 480 plus 100 is 580 plus 20 more is 600."

All three students were demonstrating their proficiency with an important tool or strategy: mental computation. Calculators are also useful math tools to use, depending on the situation. In this case we were using calculators to quickly verify answers.

I recorded *12 × 50 = 600*, and then checked the answer on the calculator.

"Let's try a bigger number, like 70," Xavier suggested.

What's the answer to 12 times 70?" I asked. I gave the students a moment and then called on Anne.

"If 12 times 50 is 600, then 12 times 60 is ten more 12s and that's 120, so 12 times 60 is 720," she explained. "So you go another 120, and you get 840, which is within the range."

Anne was using her number sense by building on what she already knew, 12 times 50, to think about 12 times 70. She was also showcasing her ability to use strategies based on place value. When Anne was finished, I recorded *70 × 12 = 840*, and then verified the answer on the calculator.

Knowing that students needed lots of modeling in order to play *Hit the Target* independently, I said to the class, "Let's try it again. This time, I'll be Player 2 and the class will be Player 1." Nicky volunteered to come up to the front of the room to record for the class and use the calculator to verify answers.

"Let's do 50," Michael suggested.

"OK, I have to think of a number to multiply by 50 and get into the target range. I'd like you to help me think of a number," I said to the class. "Tell the person next to you what you think." I made a quick tour

Teaching Tip

Calculators can be helpful tools if used appropriately and strategically.

of the room, listening to students' ideas, then called them back together. This partner talk provided an opportunity for everyone to share their ideas. After a minute or so, I asked for the students' attention and called on Michael.

"I think you should multiply 14 times 50, because in the last game we figured that 12 times 50 is 600, and we need to go higher than that," said Michael. Michael was demonstrating that he could make a reasonable guess and apply logical thinking to explore and test mathematical ideas—both important math practices.

"You're right, Michael," I acknowledged. "We need to multiply 50 by more than 12, since 12 times 50 is only 600." Michael nodded. "And your suggestion is to do 14 times 50," I added. "How much is 14 times 50? How could we figure that out mentally, class?"

"Oooh, that's hard," Gordon said.

I waited a moment and no hands were raised. I decided to model my thinking for the students to help them get their minds around the problem. "Listen to my idea and see if you can explain why it makes sense," I said. "I know that 12 times 50 is 600, that's twelve 50s. But I need two more 50s to get fourteen 50s, and two more 50s is 100 more. So 14 times 50 is 100 more than 600, and that's 700. Can someone explain my idea in your own words?" Asking students to restate someone else's idea helps them orient to the thinking of others and encourages active listening. It's also a useful assessment strategy.

"Well, to start with, we already figured out that 12 times 50 is 600," said Rebecca. "So 12 times 50 is like twelve 50s. It's like you're counting by fifties. So you keep counting up a couple more 50s, and that's 700."

"So we know that fourteen 50s won't get us into the target range," I said. Nicky recorded $14 \times 50 = 700$ on the record sheet for everyone to see and checked the answer on the calculator.

"Does anyone have an idea about where we could go from here?" I asked.

"Multiply 50 times 17, because 50 times 16 is 800, and you add another 50 to get 850," Jenny explained.

"How do you know that 50 times 16 is 800?" I asked.

"Well, in 100 there's two 50s, and split 16 in two you get 8, and 8 times 100 is 800," she reasoned, using a doubling and halving strategy (doubling the 50 and halving the 16).

"You could do 50 times 16 or 17, because both answers are within the range," observed Rebecca.

"Do you think 50 times 18 would work?" I asked.

Mathematical Practice

Making reasonable guesses and applying logical thinking are important mathematical practices (MP1 and MP3).

Teaching Tip

Can someone explain in your own words? is an important question in helping students orient to the thinking of others and actively listen.

"That's too high," answered Don. "If 50 times 17 is 850, then 50 times 18 is 900, and that's above the range."

Nicky finished the game by recording *50 × 17 = 850* and then checked the answer on the calculator.

"Now I'd like you to partner up and play *Hit the Target*," I told the class. "You and your partner need a calculator, a piece of paper, and a pencil." I then pointed to the directions for the game on the board and asked Gordon to read them aloud one more time. When he finished, the students excitedly fished their calculators out of their desks and began.

When partners had played several games of *Hit the Target*, I called the class together and had them reflect on the activity in writing using the following prompts:

- What did you like about *Hit the Target*?
- What was easy?
- What was difficult?
- What surprised you?
- What methods did you use to multiply numbers mentally?
- If you used decimals, what did you learn?

Some examples of student work are shown in Figures G3–1 through G3–4.

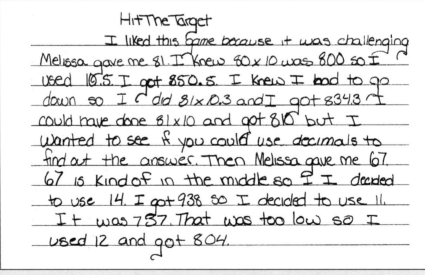

Hit The Target

I liked this game because it was challenging Melissa gave me 81. I Knew 80 x 10 was 800 so I used 10.5. I got 850.5. I knew I had to go down so I did 81 x 10.3 and I got 834.3 I could have done 81 x 10 and got 810 but I wanted to see if you could use decimals to find out the answer. Then Melissa gave me 67. 67 is Kind of in the middle so I I decided to use 14. I got 938 so I decided to use 11. It was 737. That was too low so I used 12 and got 804.

Figure G3–1. Katie used decimal numbers while playing *Hit the Target*.

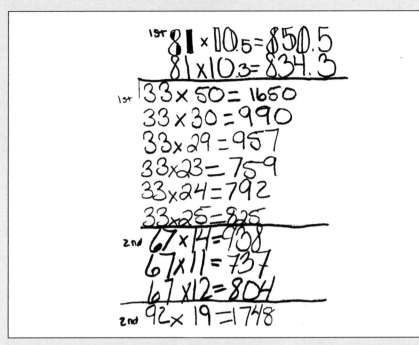

Figure G3–2. Katie and Mindy's score sheet for *Hit the Target*.

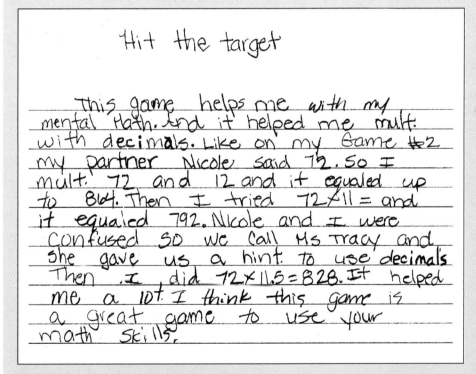

Figure G3–3. Anita noticed that *Hit the Target* helped with her mental math skills.

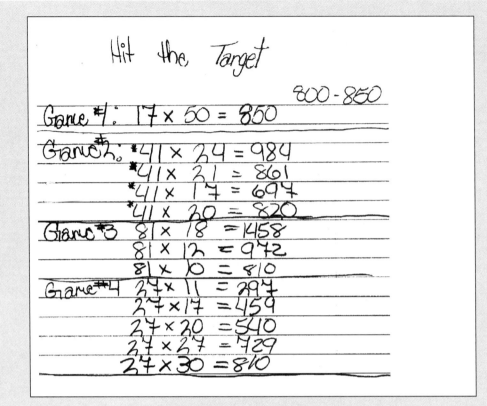

Hit the Target

800-850

Game #1: 17 × 50 = 850

Game #2: *41 × 24 = 984
        *41 × 21 = 861
        *41 × 17 = 697
        *41 × 20 = 820

Game #3 81 × 18 = 1458
         81 × 12 = 972
         81 × 10 = 810

Game #4 27 × 11 = 297
         27 × 17 = 459
         27 × 20 = 540
         27 × 27 = 729
         27 × 30 = 810

Figure G3–4. Nicole's score sheet for *Hit the Target*.

Reflecting on the Lesson

How does this game develop students' number sense?

When students rely on their intuitive reasoning about numbers and operations, they're using their number sense. The focus on estimation and mental calculation in this activity helps students develop this intuition.

After her students had learned the game, Pam Long played *Hit the Target* with them for several weeks at the beginning of math class. She changed the target range frequently and reported that the experimentation with numbers called for in the game helped improve students' estimation and mental computation skills. Most important, she said, the game required students to calculate for a purpose and to apply multiplication meaningfully and flexibly.

When students estimate while playing *Hit the Target*, they have opportunities to compare numbers and think about number relationships. Forming an estimate also involves the student in mental computation as a preliminary step. For example, if the target range is 800–850 and the starting number is 50, a game might play out like this:

$50 \times 10 = 500$ The number is too low.

$50 \times 20 = 1,000$ The number is too high.

$50 \times 15 = 750$ The number is closer but still too low.

$50 \times 17 = 850$ The number is within the target range.

With each calculation, the player is thinking about the product and comparing it to the target range. By estimating, she produces an approximate answer, one that is "close enough" to allow her to decide, *What do I do next in order to get into the target range?* While playing *Hit the Target*, students maintain oversight of their progress toward the target range, evaluating the reasonableness of their intermediate results.

Calculating mentally helps students develop their own strategies for applying operations, and helps them think flexibly. When we used the activity in Pam Long's classroom, for example, we were trying to hit the target range of 800–850 by using 75 as a starting number. Anita said, "Seventy-five times 10 is 750 and add another 75 and that's 825." Anne thought about money when solving the problem: "Pretend 75 is like three-quarters. So try to get up to 800 or 825 by counting by seventy-fives in your head. Like 75 and 75 is 150, and four 75s is 300, and so on." Michael took a completely different approach. He suggested that we divide 825 by 75 to get the number you need to multiply 75 by. All three students were solving the problem in a way that made sense to them.

What are some questions I could ask that would stimulate students' thinking about the mathematics in this game?

As a teacher, you play an important role in building students' number sense. One way to facilitate the development of number sense is to ask questions that require more than a right answer and that can prompt students to explore a mathematical idea. The following questions, which can be asked while students are playing *Hit the Target*, during a class discussion after a game, or as prompts for a writing assignment, will help you stimulate students' thinking and generate important discussions:

- What numbers were difficult to start with and which were easy? Why?
- What strategies did you use when calculating mentally? Explain your strategy.
- Did you change the target range? If so, what new target range did you use? How did the target ranges compare?

- Did you ever start with a number that was greater than the target range? If you did, explain what happened.
- How did the calculator help you in playing this game?
- Did you have to use decimal numbers in the game? If you did, explain what happened.

Get to Zero

Overview

Get to Zero gives students—individually, in pairs, or in small groups—practice with adding, subtracting, multiplying, and dividing whole numbers. Students can perform the calculations mentally or use a calculator, whichever you feel is more appropriate. Players start with a three-digit number and use any series of mathematical operations involving the numbers 1 through 9 to get to 0 in as few turns as possible.

Learning Targets

GRADE 3

I can use the four operations with whole numbers to solve problems.

GRADE 4

I can use the four operations with whole numbers to solve problems.

GRADE 5

I can use the four operations with whole numbers to solve problems.

NUMBER SENSE FEATURES

(see page xv)

Materials

A calculator for each student, pair, or group of students

⏱ Time

Forty-five minutes

133

Teaching Directions at a Glance

1. Players choose a three-digit number (example: 500).
2. Players choose an initial operation and number (example: divide 500 by 5); any number from 1 to 9 and any operation can be used. *Only whole numbers are allowed! If an operation results in a decimal answer, players must go back and try another number and/or operation.*
3. Players perform the calculation mentally or on the calculator and record the result on paper (example: $500 \div 5 = 100$), so they can look back over their work.
4. Players repeat Steps 2 and 3 until they get to 0.

Sample Game Scenario 1

Turn 1: $500 \div 5 = 100$

Turn 2: $100 \div 5 = 20$

Turn 3: $20 \div 5 = 4$

Turn 4: $4 - 4 = 0$

Sample Game Scenario 2

Turn 1: $752 \div 4 = 188$

Turn 2: $188 \div 2 = 94$

Turn 3: $94 - 4 = 90$

Turn 4: $90 \div 9 = 10$

Turn 5: $10 \div 5 = 2$

Turn 6: $2 - 2 = 0$

Extensions

1. Give everyone the same number to start with and challenge the class to get to 0 in as few operations as possible.
2. Ask students to find as many three-step numbers (those from which you can get to 0 in only three operations) as they can.

Teaching Directions with Classroom Insights

"I'm going to teach you an activity called *Get to Zero*," I began. "You'll need a calculator and a piece of paper and a pencil. You many work alone, with a partner, or with a small group."

The students in Pam Long's class had each been assigned a calculator the first week of school, and everyone now eagerly pulled it from his or her desk. Calculators can be helpful tools when solving mathematical problems. To help model the directions, I placed a calculator and a piece of paper at the front of the room where everyone could see them.

1. Players choose a three-digit number.

"To begin, you need to choose any three-digit number," I explained. "Once you've chosen your number, write it down on your paper. Would anyone like to suggest a three-digit number for us to work on together?"

"How about 500?" offered Nancy.

I wrote *500* on the paper where everyone can see it. "Our goal is to get to 0 in as few mathematical operations as possible," I said. The goal of finding shortcuts and noticing patterns naturally engages students in important mathematical practices.

2. Players choose an initial operation and number (1–9).

"You may use any operation: addition, subtraction, multiplication, or division. When using one of those operations, you may only use the numbers 1 through 9. I'll work through an example. What operation should we begin with?"

"I think you should start with division because it gets you a smaller number than all the other operations," said Carl. Students nodded their head in agreement. Although Carl's idea was obvious to the students, the comment signaled that Carl had a sense of what happens to whole numbers when you operate on them using division.

I then proposed dividing 500 by 5. I chose the number 5 because I knew it wouldn't yield an answer with a remainder. I wanted to keep things uncomplicated while modeling the activity.

Mathematical Practice

The goal of *Get to Zero* is to help students notice patterns (MP7) and to look for shortcuts when computing (MP8), two important mathematical practices.

3. Players perform the calculation mentally or on the calculator and record the result on paper.

"I'll begin by dividing 500 by 5, I stated. "What's 500 divided by 5?"

"One hundred!" the students chorused.

I then punched the numbers into my calculator, encouraging students to verify the answers using the calculators at their tables. Calculating mentally is not only often faster than a calculator and makes more sense, but also helps develop mathematical thinking.

I wrote *divide by 5* under the number *500* on my paper before I continued. This was Turn 1.

4. Players repeat Steps 2 and 3 until they get to 0.

"Now I'll divide by 5 again, I told them. "I want you to mentally divide 100 by 5."

After a few moments, several students raised their hand. I asked the class quietly to say the answer to 100 divided by 5. Then we verified the answer on the calculator to be sure it was 20. Again, I wrote *divide by 5* for Turn 2.

"Now we have 20 showing on the calculator," I said. "We want to get to 0 in as few moves as possible, so what should we do?"

"Divide by 5 again!" exclaimed Manuel.

"Or divide by 4," added Mary.

"What about divide by 10?" asked Michael.

"That wouldn't work, you can only use 1 through 9," said Todd, reminding Michael of the rules.

"Think about what would get us the smallest quotient, or answer," I suggested. After a few seconds, several hands flew up. I called on Carl.

"If you divide 20 by 5 you'll get 4, but if you divide it by 4 you'll get 5, he said. "So I think we should divide by 5."

"I don't think it matters," Gordon interjected. "Because either way you'll get to 0 in the same number of turns."

"Gordon, can you tell us more about that?" I asked.

"Well, if you divide 20 by 5, the answer's 4 and then you could subtract 4 to get to 0," he explained. "If you divide 20 by 4 you get 5 and all you have to do is subtract 5 to get to 0. For both, you get to 0 in two turns."

"Does that make sense?" I asked. Students nodded their head.

"We could choose 4 or 5 as a divisor," I said. "Let's try 5 and divide 20 by 5."

"It's 4," said Hannah.

"Okay," I responded. "In this game, it's handy to use the calculator to keep track of what's happening to the numbers as we make our way to 0." After I verified the answer on the calculator, I wrote *divide by 5* for Turn 3. "What shall we do next?" I asked.

"Subtract 4!" several students chimed in.

I finished by writing *subtract 4* for Turn 4. "So it took us four turns to get to 0," I said. "Let's try another one. Raise your hand if you have another three-digit number less than 1,000 for us to begin with."

"Let's start with an odd number," suggested Blanca. Blanca often challenged the group and she exuded confidence when working with numbers. "How about 123?"

I wrote *123* on a new piece of paper, placing it where everyone could see it. "I want you to talk with your neighbor about what operation and number we should use to begin," I said. After a moment, I asked for volunteers and called on Xavier.

"Divide 123 by 2," he suggested.

Several students groaned and others shook their head in disagreement.

"Before we divide 123 by 2, I'm interested in hearing what you think will happen," I said.

"I think the answer is going to be a number with a remainder," Cam predicted.

"Why do you think that?" I asked.

"Because 2 will go into 12 evenly, but 2 won't go into 3 evenly," he responded. Cam appeared to be solving 123 divided by 2 using the long-division algorithm.

"I think you won't end up with a whole number and you have to have a whole number for an answer or else it's really hard to get to 0," said Jenny.

"When you're dividing, how do you know if you'll get an answer that's a whole number?" I asked.

"You just have to have the feeling for what's going to divide evenly into a number, Kerry mused. "If you don't know, then you'd have to play around with the numbers to get a whole number."

"Let's use our calculators and divide 123 by 2," I instructed. Students soon realized that dividing by 2 resulted in an answer with a remainder.

"I got 61.5," Cam reported. "That's the same as 61 and a half. If you get a decimal, it's hard to get to 0."

"So it's important that you end up with a whole number for an answer whether you divide, multiply, subtract, or add. I think Xavier's

Teaching Tip
Calculators, when used appropriately and strategically, can be helpful tools when problem solving.

idea helped us learn new things about this activity," I said. "How about another idea?"

"Let's divide 123 by 3," offered Hannah. "I'm picking divide by 3 because 123 is a multiple of 3."

Students used their calculators to divide and came up with 41 as an answer. On my paper, I continued to keep track of the operations and numbers we were using, modeling for the students how they might keep track of their decisions. (Figure G4–1 shows how Niqueta kept track of her games.)

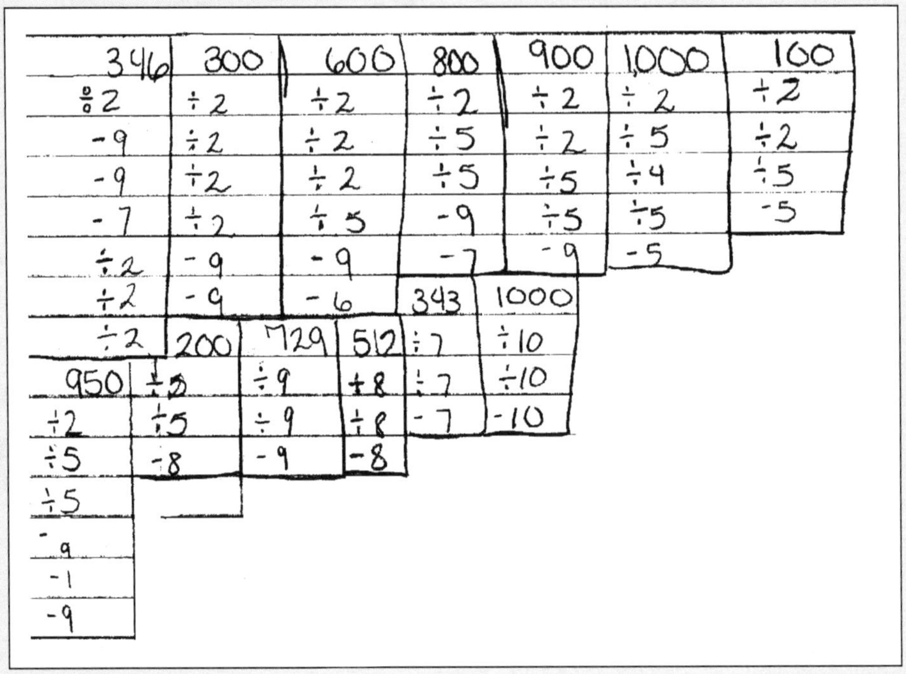

Figure G4–1. Niqueta's record of her *Get to Zero* games.

"What next?" I asked.

"We should subtract 1 so we get an even number," said Devin. "That would give us 40."

I wrote *subtract 1* on my paper. That was our second operation. Since finding shortcuts is an important math practice, I asked students to think about what to do next and reminded them that we wanted to get to 0 in as few operations as possible. Someone suggested dividing 40 by 2, but that idea was vetoed in favor of dividing by 5. Students were beginning to make sense of the game. We divided 40 by 5 to get 8, and then subtracted 8 to get to 0 in four turns.

After we finished this second game together, students got to work; some partnered up, some worked alone. The game was motivating and sustained the students' interest. I moved around the room, mostly observing and asking questions. As I observed, I noticed that being able to use calculators as tools freed the students to explore numbers and operations. They seemed uninhibited and challenged by trying to get to 0 in as few operations as possible.

After about twenty-five minutes, when it seemed that most students had explored several sequences, I asked them to write about the game (see student work samples in Figure G4–2, G4–3, and G4–4) and I provided the following prompts:

This game helps me learn . . .

I discovered . . .

I think . . .

I found out . . .

Teaching Tip

Calculators can free up students to explore numbers and operations.

Teaching Tip

Ask students to write about the game to encourage further discoveries and reflection.

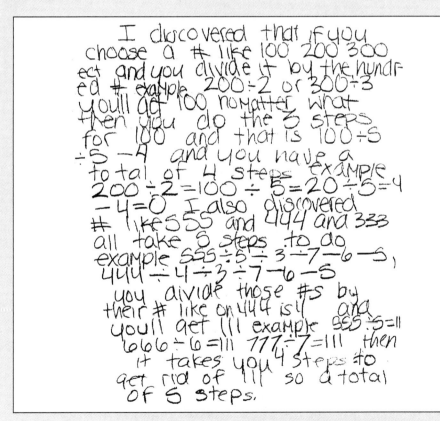

Figure G4–2. Carl discovered a pattern with certain numbers.

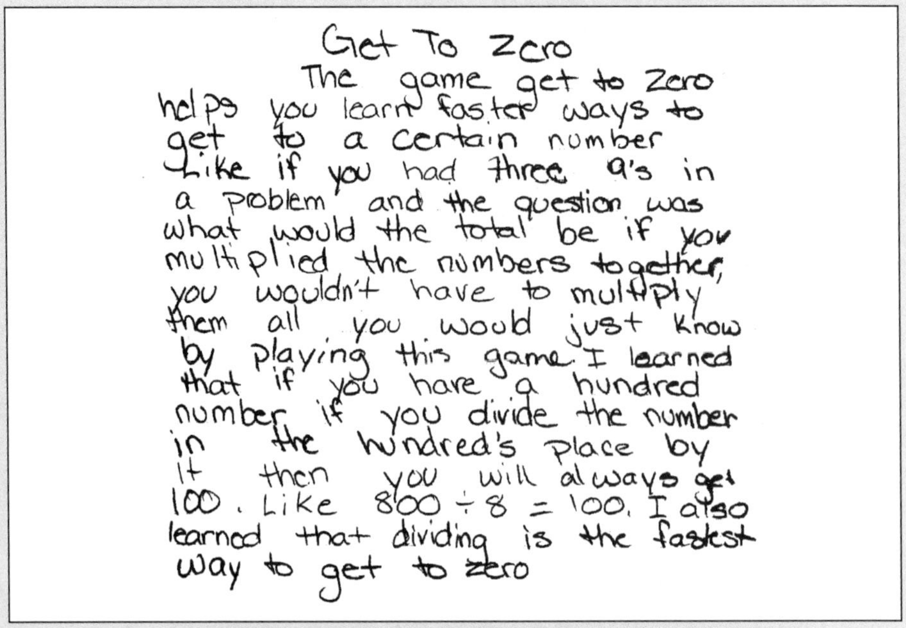

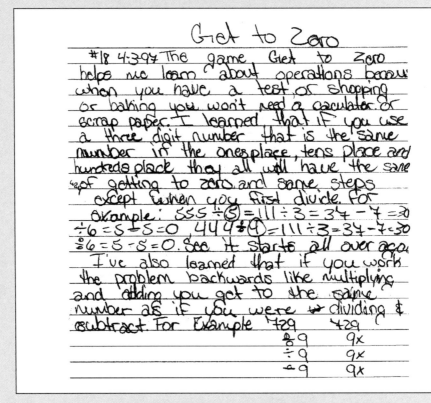

Get To Zero

The game get to zero helps you learn faster ways to get to a certain number Like if you had three 9's in a problem and the question was what would the total be if you multiplied the numbers together, you wouldn't have to multiply them all you would just know by playing this game. I learned that if you have a hundred number if you divide the number in the hundred's place by it then you will always get 100. Like 800 ÷ 8 = 100. I also learned that dividing is the fastest way to get to zero

Figure G4–3. Katie realized that dividing is the fastest way to get to 0. She also discovered how to get to 100 quickly.

Get to Zero

#18 4·3·97 The game Get to Zero helps me learn about operations becau' when you have a test, or shopping or baking you won't need a caculator or scrap paper. I learned that if you use a three digit number that is the same number in the onesplace, tens place and hundreds plack they all will have the same of getting to zero and same steps except when you first divide. For example: 555 ÷ 5 = 111 ÷ 3 = 37 − 7 = 30 ÷ 6 = 5 − 5 = 0   444 ÷ 4 = 111 ÷ 3 = 37 − 7 = 30 ÷ 6 = 5 − 5 = 0. See it starts all over aga. I've also learned that if you work the problem backwards like multiplying and adding you get to the same number as if you were to dividing & subtract For Example 729 / 729

	÷ 9	9x
	÷ 9	9x
	− 9	9x

Figure G4–4. Jenny noticed that *Get to Zero* helped her learn about operations. She also explained how to get from 729 to 0 in three moves.

Reflecting on the Lesson

How does this activity help students develop their number sense?

Facility with computing is an important characteristic of number sense. *Get to Zero* gives students practice with all the operations, especially division. And using calculators allows them to take risks and try new ways of thinking.

This activity gives students opportunities to think about operations and what happens to quantities when they're added, subtracted, multiplied, and divided. During the activity, for example, Carl revealed his knowledge about division in relation to other operations when he commented, "you should start with division because it gets you a smaller number." When Anne recommended subtracting until your number ends in 0 so that you can divide it by 5 evenly, she was drawing on her understanding of the multiples of 5 and of division. She was using her number sense to think about a strategy.

Get to Zero also gives students the chance to explore the characteristics of numbers: odd and even, factors, multiples, prime, and composite. In addition, students are able to learn about decimal numbers, look for patterns, and make conjectures and test hypotheses, all of which help develop number sense.

How can *Get to Zero* be modified so that it's more easily accessible?

One way to make this activity more accessible is to start with numbers 50 and 100 instead of any three-digit number and to use the digits 1 through 5 instead of 1 through 9. Working with smaller numbers is less daunting for students and is therefore a more comfortable place to start.

A colleague of mine taught *Get to Zero* to her class using only the numbers 50 through 100 and the digits 1 through 5. Then she asked her student to describe strategies they thought were helpful in getting to 0. Here are the strategies they came up with:

1. Choose even numbers.
2. Try to divide by the biggest number you can.
3. If you can't divide, then add or subtract to get a number that ends in 5 or 0.
4. Never divide when you have an odd number.
5. If you have an odd number, subtract 1, 3, or 5 to get an even number.

How would you help students become aware of the divisibility rules?

While some students discover rules for divisibility on their own, it helps to make them explicit so that all students have access to them. Still, students will need a good deal of experience to become comfortable with the divisibility rules. Encourage them to look for patterns.

Some are easy to recognize, such as counting by fives and noticing that all the multiples end in 0 or 5, or learning that all even numbers are multiples of 2. But the discovery that if the sum of the digits in a number is a multiple of 3 (for example, 123, or $1 + 2 + 3 = 6$), then the statement "the number is a multiple of 3" is not as obvious.

The benefit of a game like *Get to Zero* is that it provides a reason for students to think about divisibility, which is another valuable way to understand the relationships among numbers and a key aspect of number sense.

Get to 1,000 (Addition)

Overview

Get to 1,000 (Addition) is essentially the same activity as *Get to 1,000 (Multiplication)*, with a slight twist so that the focus is on place value and addition rather than on multiplication. On each roll of the die, players decide whether to make the number ones, tens, or hundreds (e.g., if the player rolls a 3, she can choose to keep the number as 3, or make it 30, or 300). Players keep a running total during the game and figure their final score after ten rolls. The winner is the player who is closest to 1,000 (the total may be under or over 1,000).

NUMBER SENSE FEATURES

(see page xv)

Learning Targets

GRADE 3

I can quickly and easily add and subtract numbers within 1,000.

GRADE 4

I can add and subtract larger numbers.

Materials

One die for each pair of students
One score sheet (see Reproducible 2 for a blank score sheet) for each student

 ## Time

Forty-five minutes

To download the Reproducible, please visit www.mathsolutions.com/ mathworkshopessentialsnumbersense.

Teaching Directions at a Glance

1. Players, in pairs, take turns rolling a die. On each roll, a player decides whether he or she wants to make the number on the die ones, or tens, or hundreds. For example, if the player rolls a 3, she can make a 3, or 30, or 300. After each roll, players record the number on their score sheet.

2. Players continue to roll; decide whether to make the number ones, or tens, or hundreds; and record the number until they each have ten numbers on their score sheet. (Each player will have rolled the die ten times.)

3. Each player finds the sum of his or her ten numbers. The one whose final score is closer to 1,000, whether over or under, is the winner.

Teaching Directions with Classroom Insights

From a Third-Grade Classroom

I began the lesson by showing the class an enlarged version of the score sheet for *Get to 1,000 (Addition)* that I had posted on the board:

Total

Total

"Today, we're going to play a math game called *Get to 1,000, Addition*," I began, pointing to the score sheet. "The goal is to get as close to 1,000 as possible—closer than your partner. When you play, you and your partner will each have a score sheet and take turns rolling the die. But first, I'm going to model the game for you, so you know how to play."

1. Players, in pairs, take turns rolling a die. On each roll, a player decides whether he or she wants to make the number on the die ones, or tens, or hundreds.

I rolled the die and got a 6. "I rolled a 6. Now I have to decide whether to make it ones, or tens, or hundreds. If I make it ones, I record *6* in the first box on my record sheet. If I make it tens, I record *60*. If I make it hundreds, I record . . . " I prompted. I wanted to see if students were getting the gist. In a choral voice, many of the students shouted, "*Six hundred!*"

I then placed my index finger on my chin, indicating to the class that I was thinking about what to do. "I think I'll make the roll on the die tens, so I'll write down the number *60* in the first box on my score sheet." Now the score sheet looked like this:

Total

60									

2. Players continue to roll; decide whether to make the number ones, or tens, or hundreds; and record the number until they each have ten numbers on their score sheet.

"When you play, you'll play with a partner and take turns rolling the die and recording on your own score sheet," I explained. I rolled the die again and got a 2. "What advice do you have for me?" I asked the class. Asking questions while I'm modeling a game helps to keep everyone engaged, and it gives students a chance to think mathematically in ways that can develop their number sense.

Annalise responded, "I think you should make it 200 because you're still far away from 1,000."

It's important that students justify their reasoning because sense making and reasoning are intertwined.

By sharing their thinking, students can cement their understanding, work through misconceptions, and also help further other students' understanding. Sentence frames or prompts can help students engage in justifying their reasoning:

> I think _____. I think this because _____.

Teaching Tip

Ask students to justify their reasoning; this is important because sense making and reasoning are intertwined.

I recorded Analise's idea on the score sheet, added 60 and 200 aloud, and jotted down the sum underneath the second box to keep track of my score:

Total

60	200								

260

"You can keep track of your score by adding together the numbers in the boxes and writing the total underneath," I suggested. "You can add the numbers together mentally or with paper and pencil."

I rolled again and got a 5. "What should I do now?" I asked the class. Dylan suggested that I "make it tens 'cause if you make it hundreds you'll be really close to one thousand." I listened to his advice and recorded *50* in the box on the score sheet.

I continued to model the game by rolling the die, listening to students' suggestions, and recording the numbers on the score sheet. After five rolls of the die, the score sheet looked like this:

Total

60	200	50	100	6					

260   310   410   416

"We're halfway to the end of the game, and we have 416 points," I said. Talk with a partner about how far away from 1,000 we are and what you want to do on the next roll."

As partners discussed my question, I made a quick lap around the room, listening in on their conversations. It was interesting how different students figured the distance between 416 and 1,000. Some were writing equations on paper, subtracting 416 from 1,000. Others were adding up from 416 out loud by counting up by hundreds. And others were using their number sense, rounding 416 to 400 and then working from there. When I called them back together, I had a few different students share their thinking and what they'd do on the next roll.

"I would make the next number hundreds, unless it is a high number like 5 or 6," Kristina shared. "If it's a high number, that would put us close to or over 1,000."

"I agree with Kristina," Jimmy said.

"I would keep making the numbers tens or ones to stay safe," Haroon added.

"I think you should make hundreds and get close to 1,000, and then do tens or ones," James said.

"These are good examples of strategies," I told the class. "When you play against your partner, try to think about strategies for winning the game." Discussing strategies is key in this game, and an important math practice, because it can lead students to think about the reasonableness of their decisions and intermediate results as they make their way to 1,000.

Teaching Tip

Ask students to discuss their strategies; this can lead them to reflect on the reasonableness of their decisions in the game.

3. Each player finds the sum of his or her ten numbers.

When I finished modeling the game and totaled my score, students partnered up, and I passed out the *Get to 1,000 (Addition)* score sheet (see Reproducible 2).

For a downloadable version of R-2, please visit www.mathsolutions.com/mathworkshopessentialsnumbersense.

As I made my way from group to group, I observed a wide variety of problem solving and strategic thinking. When I visited Analise and her partner, Analise was figuring the answer to 340 + 200 mentally. She quickly wrote down *540*.

"How did you know that so quickly?" I asked her.

"Well, it's easy," she responded. "'Cause 300 plus 200 is 500. Then you just have to add 40."

Analise was making use of her understanding of our base ten place value system by decomposing 340 into 300 plus 40. Understanding place value and using number decomposition and mental math are important aspects of having number sense.

Elizabeth was on her fifth roll when I stopped by her table. Her score was 580 and she had rolled a 5 on her sixth roll. While she was thinking of what decision to make, I asked, "What if you make the 5 a 500, would that be smart?"

"No way!" she exclaimed. "That would put me over 1,000 right away!"

I smiled, knowing that Elizabeth had a good idea of what happens to numbers when we operate on them. She also had a sense of how far away she was from the target number.

In contrast to Elizabeth, Layla was figuring the answer to 810 + 90 when I visited her table. Rather than solve the problem mentally by using what she knew about combining quantities that make friendly numbers (e.g., 90 + 10 = 100), Layla chose to use the standard algorithm to find the

Formative Assessment

Observe the students. What strategies are students using?

Formative Assessment

Observe students; are they breaking numbers apart in order to make computations easier? This is a foundation of number sense.

Teaching Tip

Gather the class together for a whole-class discussion during which they can share their strategies.

answer. While using standard procedures with paper and paper is fine, students with number sense often use strategies that make use of place value, number decomposition, and properties of operations. Once Layla finished finding the answer, I asked her if she could think of another way to solve the problem. She seemed stumped, so I asked her a question that was intended to help her solve the problem mentally.

"Layla, let's think about the problem 810 + 90," I began. "Is there a number inside 810 that you might add to 90 to make 100?" Layla still seemed confused.

"What's 90 plus 10?" I asked, pointing to the 10 inside 810. Layla's face lit up. "Oh! Ninety and 10 is 100! Then you have 900!"

When it seemed that everyone had the time to play at least two games, I asked for their attention and finished the lesson with a class discussion that focused on students' strategies for winning the game.

Before having a whole-class discussion, I had the students think independently about the strategies they used to win the game, and then share with their partner. Individual think time and pair-shares give students time to rehearse before a class discussion. This practice is especially helpful for English language learners.

After a couple of minutes, I had a few students share their thoughts. I called on Elizabeth first.

"I should play fun games like this one at home, especially when I'm bored." Lots of students nodded in agreement.

I agreed. "Yes, it's pretty fun and engaging. Who can share with us a strategy you used to win the game?"

"I learned that I have to think about place value and I had to use addition," Nathan said. "My strategy is that when you get at least two hundreds you turn the numbers into tens or ones."

"Nod your head if you started with hundreds and then changed to making the numbers tens and ones?" I asked, encouraging students to make connections and stay engaged.

"My strategy was if the total on the die was 1 or 2 or sometimes 3, it'd be hundreds," Dylan explained. "But if the die came up 4 or 5, I'd mostly make it tens, and if the die came up 6, I'd make tens or ones."

"My strategy was to mix the numbers," added Ella.

"What do you mean by 'mix the numbers'?" I probed.

"Like I would sometimes make the numbers on the die ones, tens, and sometimes hundreds—mix it up."

Pauline went next. "I learned that if you have a high number you use ones and tens."

"I kind of agree with Pauline," Judy said. "If you start with hundreds you will get really close to 1,000. When you get close, you start to do tens and ones."

After taking a few more comments, I had the students take out their math journals and write about their strategies (see Figures G5–1 and G5–2). Writing in math can help students extend their thinking and cement their understanding. It can also serve as an assessment artifact. Before they wrote in their journals, I provided the following prompt:

Today I played *Get to 1,000*. I learned _____.

A strategy I used was _____.

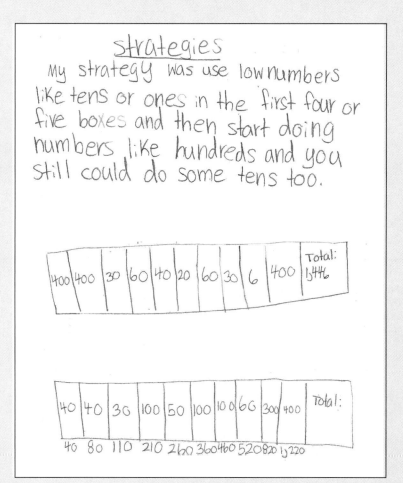

Figure G5–1. Kristina explains her strategy and shows one of the games she played. Notice how she kept track of her running totals.

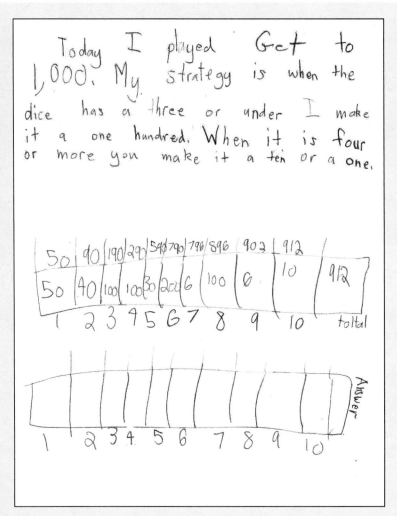

Today I played Get to 1,000. My strategy is when the dice has a three or under I make it a one hundred. When it is four or more you make it a ten or a one.

Figure G5–2. Jami's strategy was to start by making the dice rolls that were three or under into one hundred.

Reflecting on the Lesson

What were the most important benefits to students as a result of playing Get to 1,000 (Addition)?

Students benefit from playing the game in several ways. *Get to 1,000 (Addition)* helps students develop fluency with adding large numbers. Procedural fluency is an important part of having number sense. Because students are encouraged to keep a mental running total of their scores, we noticed that many were adding the numbers together using strategies based on place value. For example, when Elizabeth solved 810 + 90, she combined 90 and 10 to make 100, and then added the hundred she made to 800. We saw a lot of students use their number sense when adding during the game. Another example is when Jun used skip-counting by hundreds when adding 400 to 1,321. He could have used paper and pencil and the standard algorithm to figure the answer, but he used a far more efficient strategy guided by his number sense.

Another benefit of the game is that it requires students to use strategic thinking and communicate that thinking to others. Students get practice constructing arguments and critiquing the arguments of others when they have to communicate their strategies. Furthermore, when we ask students to write about their strategies, it helps them cement their understanding. The process of writing helps students reflect on what they know and understand.

How does this game help students develop their number sense and engage in important mathematical practices?

This game helps students develop their number sense in so many ways! The game gives them practice with mental computation. It provides opportunities to round to friendly numbers when adding. Students must consider place value and the effects of operations on numbers. And it gives students a chance to use strategic thinking, reasoning about how far away a number is from 1,000 and what to do to get close to the target number. The game also challenges the students to work within given constraints and goals in order to win the game. For example, a player has a target number to reach and has limited choices when making decisions about how to reach the target.

Isn't *Get to 1,000 (Addition)* the same as *Get to 1,000 (Multiplication)*? If so, why would I want to teach both games to my students?

Yes, the two games are essentially the same activity, but with a slight twist. *Get to 1,000 (Addition)* focuses students' attention on place value, while *Get to 1,000 (Multiplication)* has students think about what happens to numbers when you multiply them by 1, 10, and 100. I would teach the addition version first, followed by the multiplication version of the game, and then ask students how the games are alike. The ensuing discussion can help them see that in our base ten number system, each place value is ten times greater than the digit to the right, and ten times less than the digit to the left.

Get to 1,000 (Multiplication)

Overview

Get to 1,000 (Multiplication) is a two-person game that gives students practice with multiplying by powers of 10 and with addition. Players multiply the number that comes up on each of ten rolls of the die by 1, 10, or 100. Then they add the ten products. The total may be under or over 1,000. The player whose final score is closer to 1,000 is the winner. Variations of the game give students practice multiplying by 5, 25, and 50.

Learning Targets

GRADE 3

I can multiply any one-digit whole number by a multiple of 10 (6 × 90, 4 × 30).

I can quickly and easily add and subtract numbers within 1,000.

GRADE 4

I can multiply a whole number up to four digits by a one-digit number.

I can add and subtract larger numbers.

GRADE 5

I can easily multiply whole numbers.

Materials

One die for each pair of students
One score sheet (see Reproducible 3 for a blank score sheet) for each student

NUMBER SENSE FEATURES

(see page xv)

To download the Reproducible, please visit www.mathsolutions.com/mathworkshopessentialsnumbersense.

153

 **Time**

Forty-five minutes

Teaching Directions at a Glance

1. Players, in pairs, take turns rolling a die. On each roll, each player decides whether he or she wants to multiply the number on the die by 1, 10, or 100.

2. Each player records the resulting product on their score sheet.

3. Players continue to roll, multiply, and record until they each have ten products. (Each player will have rolled the die five times.)

4. Each player finds the sum of his or her products to see who is closest to 1,000.

Extensions

1. Version A: roll the die ten times and multiply the number on the die by 1, 5, or 50.

2. Version B: roll the die ten times and multiply the number on the die by 10, 25, or 50.

3. Version C: roll the die seven times and multiply the number on the die by 1, 10, or 100.

Teaching Directions with Classroom Insights

From a Fourth-Grade Classroom

I told Kathleen Gallagher's class I wanted to teach them a game called *Get to 1,000* that involved multiplication, addition, and subtraction. I explained that after I showed them how, they'd be able to play it with a partner.

"For this game, you want to get as close to 1,000 as possible," I began. "You may go over or under 1,000, but you want to get as close as you can. Each time you roll the die, you have a choice whether to multiply the number on the die by 1, 10, or 100." I wrote $\times 1$, $\times 10$, and $\times 100$ on the board. "Each time I get another product, I'll write it down on my *Get to 1,000* score sheet (see Reproducible 3 for a blank score sheet). I placed a *Get to 1,000* score sheet in front of the classroom where everyone could see it:

Total

Total

"This is a game that you play against a partner, so I have a score sheet and my partner will also have a score sheet. We'll take turns rolling the die. As we play the game, we'll each try to keep track of our scores in our head, so that we know about how close we are to 1,000. After ten rolls, we'll each add up the products and see who got closer to a score of 1,000." I noticed some blank looks. "We'll play a practice game—I'll play against the class."

> For a downloadable version of R-3, please visit www.mathsolutions.com/ mathworkshopessentialsnumbersense.

1. Players, in pairs, take turns rolling a die. On each roll, each player decides whether he or she wants to multiply the number on the die by 1, 10, or 100.

I rolled the die and got a 4. "I can multiply four by 1, 10, or 100," I said, thinking out loud. "I'm going to choose to multiply 4 by 100. What's 4 times 100?"

"Four hundred!" the class chorused.

"How do you know that?" I asked.

"Because you take 4 and add two 0s," said Elias. "Four times 100 means you do 4 one hundred times."

"It's like 400 pennies," added Brad.

"Look, you skip-count by hundreds," Nadine pointed out matter-of-factly.

"Lets' try that together and see if it works," I suggested. Together we counted by hundreds to 400. As we counted, I realized that for a great many children in the class, the answer to 4×100 was obvious. I also knew that this part of our discussion could give some of them different ways to think about multiplication. The mathematical understanding of the students in the room varied from solid to extremely fragile. Skip-counting by hundreds not only gave everyone practice with a strategy, but it also engaged everyone.

2. Each player records the resulting product on their score sheet.

I wrote *400* in the first box on my score sheet and asked for a volunteer to come up and play for the class. Ryanna came up to the front of the room, rolled the die, and announced she got a 5.

"Hmm," Ryanna voiced as she put her finger to her chin, thinking about what to multiply by 5. She finally decided to multiply 5 by 10 and recorded *50* in the first box on the class's score sheet.

3. Players continue to roll, multiply, and record until they each have ten products.

Now it was my turn. I rolled and got a 6. "I'm going to multiply 6 times 10 this time," I said. "Can someone explain why that might be a good choice?" A lot of hands popped up, and I called on Cathy.

"Well, if you multiplied 6 times 100 you'd get 600, and that would put you at 1,000 already," she explained. "So 6 times 10 is a good move. I think you could multiply 6 times 1 and that would also be a good move." Cathy was engaged in two important math practices: constructing an argument for why my move was a good one and using the big picture (making sure I didn't go over 1,000) to evaluate the reasonableness of possible decisions I could make.

Being able to examine the reasonableness of answers is both a sign of number sense and a good test-taking skill.

"What's 6 times 10?" I asked the class.

"Sixty!" they said together.

"Can someone explain how you know 6 times 10 is 60?" I asked. I wanted the students to know that I was interested in more than correct answers. I wanted them to explain their ideas in order to increase the

Mathematical Practice

Two important practices to engage students in are constructing arguments (MP3) and evaluating the reasonableness of decisions they make while problem solving (MP2).

learning potential and shift the focus from reporting answers to thinking about strategies for finding answers.

"Because 10 six times is 60," said Sharon.

"You can count by 10 six times," explained Jim.

"It's like six dimes," offered Simon.

I wrote *60* on my score sheet in the box next to the *400*.

It was the class's turn, so I called on a volunteer to come up to the front. Nick rolled for the class and got a 3. He multiplied 3 by 100 and recorded *300* in the second box. Now our score sheets looked like this:

Total

400	60									

Total

50	300									

Pointing to my score sheet, I asked the class, "So how much do I have so far?" It's important for students to know that they need to keep a running total of the score, finding the sums of the numbers in the boxes on their score sheet. Students can keep a running total by finding the sums mentally, or by using paper and pencil. I encourage students to try and figure out their running total mentally because it helps develop their number sense.

When we figured that I had a total of 460 and the class had a total of 350, I continued modeling the game.

It was now my turn, so I rolled the die and got another 6. "What should I do now?" I asked. "Any advice?" Asking students for advice keeps everyone engaged and thinking. Sometimes I'll have students talk to a partner before giving me advice, but this time I called on a volunteer right away, keeping my pacing of the lesson in mind.

"I think you should multiply by 10 because you already have over 400, and if you multiply by 100, you'll go over 1,000," Sue explained, justifying her reasoning.

"Does that make sense?" I asked. Students nodded their head in agreement, and I added another *60* to the list of figures on my score sheet.

Teaching Tip

Have students talk about their strategies rather than just report answers; this shifts the focus to number sense.

Marcos played for the class next and rolled a 2. He multiplied 2×10 and wrote *20* on the class's score sheet.

The students and I took turns rolling the die, choosing multipliers, and recording the results a few more times. After five rolls each, our score sheets looked like this:

Total

400	60	60	100	10					

Total

50	300	20	50	6					

We had just finished our fifth roll and were halfway through the game when I asked the students to talk with a partner about three questions:

How much do we each have so far?

How far away are we from 1,000?

Can you think of a strategy that will help you win?

These questions gave students practice keeping a running total of the numbers on their score sheets, and it gave them an opportunity to think about strategy—how far away from 1,000 am I, and what multipliers do I need to use with which numbers in order to get close to 1,000? This kind of reasoning requires students to use their number sense and evaluate the reasonableness of their intermediate results, an important math practice.

4. Each player finds the sum of his or her products to see who is closest to 1,000.

After finding the sum of each of our ten products, students partnered up to play *Get to 1,000 (Multiplication)*. Before sending them off to play, I posted a few sentence frames and prompts to help students communicate with their partner during the game:

My score is _____.

I am _____ away from 1,000, or, I need _____ more to make 1,000.

Mathematical Practice

Ask students to evaluate the reasonableness of their results (MP8); this helps them focus on number sense.

I rolled a _____ and multiplied it by _____ because _____

_____ .

My strategy for winning the game is _____

_____ .

As students played I walked from table to table, listening in on conversations, posing questions, and encouraging the students to use the frames and prompts to communicate their thinking. Rob and Charles were on their final roll when I joined their game.

"I've got 800. I'm getting close," Charles said.

He was thinking out loud and was ready to make the final roll when I stopped him with my question: "How far away from 1,000 are you and what would have to happen for you to make 1,000 exactly?"

Charles thought for a while. Rob was dying to answer, but I put up my hand, signaling him to be patient and give Charles a chance.

"Hmm. I need 200 more to make 1,000. If I rolled a 1 and multiplied by 100, I'd get to 900, so If I rolled a two and multiplied it by 100, I'd get there exactly," he reasoned.

"Do you think it's likely that someone would get 1,000 exactly?" I asked.

Charles and Rob looked at each other, not quite sure how to respond. Leaving them to think about this question, I moved on to another table.

As the students explored the game, my main responsibility was to ask them questions that would prompt them to explain their reasoning and use their number sense. What's your score so far? How far away from 1,000 are you and how do you know? What strategy are you using to win the game? Who's winning and how do you know? How did you figure out the answer when you used that multiplier? All of these questions engage students and keep them focused on the math while playing *Get to 1,000 (Multiplication)*.

When students were finished playing several games, I had them write about and then share the strategy they used to win the game.

Jenny wrote: *All I did was roll the dice. I placed my numbers down with what made sense. Like if I had 933 and got a 6 on my last roll I would put it as 60.*

Kimm wrote (see Figure G6–1 on the next page): *For [my] number one [game], I choose those numbers because I thought it would help me. But it didn't. What I should of done was put more hundreds than 10's because I think that would of helped me. For [my] number two [game], I got a little better with it because I put more hundreds than more tens and . . . keeped the answer [total] in my head . . .*

Teaching Tip

Observe the students. Pose questions and encourage students to communicate their thinking.

Carl wrote: *When playing* Get to 1,000, *I learned you have to ceep track of the numbers because you mite pass 1,000. I liked playing because I can learn how to add better. In the first game I got 970 and in the second game I got 980. In the first game I needed 30 to get to 1,000 in the second game I needed 20 to get to 1,000.*

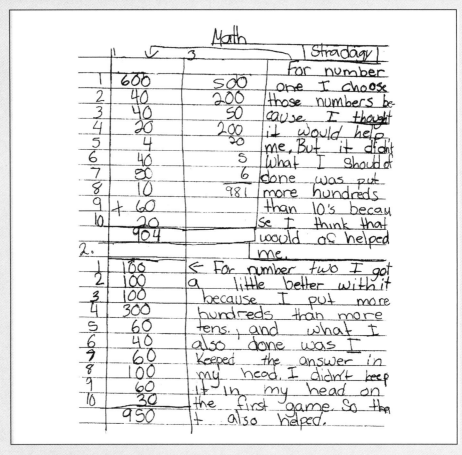

Figure G6–1. Kimm improved her second game because of what she learned during her first game.

Reflecting on the Lesson

What is the purpose of this activity?

In this activity, we want students to think about numbers, their magnitude, and their relationship to 1,000. We also want students to get a sense of what multiplication does to numbers. These are important characteristics of number sense.

One of the things that we like about this game is that students must make decisions based on how close or how far away they are from 1,000. In other words, they must think about numbers and how those numbers relate to 1,000.

For example, when I was modeling the game for Kathleen Gallagher's students, I had a score of 460 and rolled a 6. When Sue responded, "I think you should multiply by 10 because you already have over 400 and if you multiply by 100 you'll go over 1,000," she was thinking about the effect multiplication has on numbers and how close 460 is from 1,000. Sue was reasoning about numbers, which is what students must do in order to develop their number sense.

Another focus of the activity is for students to experience and become proficient at multiplying by multiples of 10 and other important landmark numbers, like 25. Because our number system is based on powers of 10, the numbers 10, 100, 1,000, and their multiples are especially important landmark numbers.

Landmark numbers are familiar landing places that make for simple calculations and to which other numbers can be related. In solving problems, people with well-developed number sense draw on their knowledge of these important landmarks.

Finally, the game also challenges the students to work within given constraints and goals in order to win the game. For example, a player has a target number to reach and has limited choices when making decisions about how to reach the target.

If I use *Get to 1,000 (Multiplication)*, I'll need to manage a classroom in which students are active, vocal, and working in groups. What advice do you have?

It isn't easy to manage a classroom, and in some ways it's even more challenging to manage an active one. The best-laid plans can still lead to disaster if students aren't listening and following directions.

What's most important is that you work to establish a classroom environment in which mutual respect is valued and modeled. Time set aside for practicing how to use materials, how to take turns, how to communicate with others—including active listening and respectfully disagreeing—is time well spent.

Giving clear directions and modeling effectively are also important ingredients in managing a math class. Children who know what's expected of them are less likely to misbehave.

How can I assess a student's number sense during *Get to 1,000 (Multiplication)*?

When we use this activity, we have several questions in mind as we watch students play the game and as we listen to their questions and responses during class discussions:

- Are students able to multiply numbers by 10 and multiples of ten? Are they able to multiply numbers by 25? by other landmark numbers?
- Do students know the affect multiplication has on numbers? Do they have a sense of how big a number will become when you multiply it by 1, 10, or 100?
- Do students have a sense of how close they are to 1,000 when they're playing the game?
- Are students using strategies to win the game? What strategies are they using and are they able to articulate them?
- Are students able to keep track of their score mentally or do they use paper and pencil?
- Are students accurate in their computations?
- How do students figure out the difference between their score and their partner's score?
- Can students figure out the difference mentally?

Can I use this game in a third-grade classroom?

Absolutely! We've taught *Get to 1,000 (Multiplication)* in the third grade and it has been very effective, especially toward the last half of the year when third graders have developed an understanding of multiplication. To differentiate the game, we have adapted the rules so that rather than having 1,000 as the goal, we've asked some third graders to try to get to 100, using 1 through 6 as multipliers. What's nice about this game is that it can easily be adapted to the skill level of the players by adjusting the goal, the number of rolls of the die, and the multipliers. See Game 5: *Get to 1,000: Addition* for another version of the game. No matter what version of the game is played, the important thing is that students think about the affect multiplication (or addition) has on numbers.

GAME

Decimal Maze

Overview

Numbers and operations in base ten is a key domain for elementary students and decimal numbers are a fundamental aspect of our base ten number system. It's important for upper-grade students to see the connections rather than focus on computational procedures. *Decimal Maze* offers students a context for exploring and considering what happens when we use the four operations with decimal numbers. Starting with whole-number situations and then moving to decimal numbers, students build on their prior understandings and expand them to make sense of the base ten system when the numbers are less than 1.

Learning Targets

GRADE 4

I can compare two decimals to the hundredths.

I can use models to justify decimal comparisons.

GRADE 5

I can read, write, and compare decimals to the thousandths.

I can compare two decimal numbers.

I can use place value understanding to round decimals.

I can add, subtract, multiply, and divide decimals to the hundredths using my place value understanding.

Materials

Decimal Maze game boards (see Reproducible 4)
Decimal Maze recording sheets, if desired
Cubes or other small tokens
Calculators or devices with calculators

NUMBER SENSE FEATURES

(see page xv)

To download the Reproducible, please visit www.mathsolutions.com/ mathworkshopessentialsnumbersense.

 **Time**

Forty-five minutes to one hour

Teaching Directions at a Glance

1. Begin with a series of number talks relating whole-number multiplication and division to operating with decimals.

2. Show the *Decimal Maze* game board and model part of a game with the class.

3. Begin with a value of 100 on your calculator. As you cross a segment, perform the indicated operation on your calculator.

4. Move down or sideways (never up) through the maze from start to finish. You may not retrace any steps.

5. The goal is to choose a path that results in the largest value when you reach the finish.

6. Ask students to pay attention to their strategies and what they notice about operating on decimal numbers as they play the game.

7. Go back to the original problems from the number talk and ask students to add to what they know.

Teaching Directions with Classroom Insights

From a Fifth-Grade Classroom

I introduced myself to Lynn Lorimer's fifth graders and then wrote some sentence frames on the board:

The answer will be about _____ because _____.

It's more/less than _____ because _____.

It's between _____ and _____ because _____.

"OK," I told the class, "I'm going to write a problem on the board. Your job is to think and estimate. I'm not asking for the exact answer. The job here is to pay attention to how your brain thinks about the numbers and operations to make a reasonable estimate. I'll ask you how you came up with your estimate, and you can use these sentence frames if you'd like."

1. Begin with a series of number talks relating whole-number multiplication and division to operating with decimals.

I wrote the first problem on the board:

$$27 \times 3 =$$

Students had a variety of estimation strategies. I encouraged them to use the sentence frames to describe their thinking. Some students decomposed the 27 into a 20 and a 7; others rounded to friendly numbers like 25 or 30. As students described their thinking I named the strategies they used, hoping they'd apply them to the next problem with a decimal number.

I gave them the next problem, saying, "Six times $\frac{9}{10}$" as I wrote it on the board.

$$6 \times 0.9 =$$

Some students immediately saw that 0.9 was very close to 1 so the answer should be close to 6. Other students were thrown and fell back on rules and procedures. These students had a hard time estimating and instead wanted to tell the exact answer. They also seemed to have less number sense. One girl explained that the answer was 5.4 because "you multiply and then move the decimal."

Rather than get into a lengthy discussion, I acknowledged that there were a variety of strategies students were using and a variety of answers students were coming up with. We moved on to a pair of division problems using the same protocol.

$$112 \div 2 =$$
$$15 \div 0.5 =$$

We had a fascinating discussion about what the second problem meant. I had students talk to a partner about whether the answer would be more than 15 or less than 15. Students who had number and operation sense strategies, such as looking for friendly numbers (e.g., 112 is close to

Teaching Tip

Let students know in advance that you are going to ask about their thinking; this gets them to focus on their solution strategy and not just getting the answer.

Teaching Tip

It's very helpful for students to name and discuss decimal numbers in terms of their fractional parts. Decimal numbers are fractions and students can tap into their experiences with fractions to make sense of decimals.

100) or decomposing numbers (15 is the same as a 10 and a 5), had more success in estimating and recognizing reasonable answers. The students who were stuck in procedural thinking lost the "big picture" and seemed unperturbed if their answers were off by a magnitude of 10 or even 100.

Natalie remarked that she hadn't learned how to compute $112 \div 2$ in her head. I asked her to think about what the equation means. Someone suggested it could be read as, *how many 2s in 112?* This precise language not only helped with the whole-number division problem, it also gave students insights into the division with a decimal. If $112 \div 2$ means *how many 2s in 112*, then $15 \div 0.5$ means *how many 0.5s (or $\frac{1}{2}$s) are there in 15?* This perspective also helped students see why dividing by a number less than one leads to a larger quotient.

We didn't belabor the number talks. My intention was to float some strategies and questions that students would continue to consider as they engaged in the *Decimal Maze* game.

REPRODUCIBLE 4

NAME _____

Decimal Maze Game Board

Move down or sideways (never up) through the maze from *Start* to *Finish*. You may not retrace any steps.

Begin with a value of 100 on your calculator. As you cross a segment, perform the indicated operation on your calculator.

The goal is to choose a path that results in the largest value when you reach *Finish*.

For a downloadable version of R-4, please visit www.mathsolutions.com/mathworkshopessentialsnumbersense.

2. Show the *Decimal Maze* game board and model part of a game with the class.

I quickly modeled part of the *Decimal Maze* game with the whole class. I put two different-colored centimeter cubes at the top of the game board on 100 and offered to go first since I knew how to play. (See Reproducible 4 for the game board.)

"So each play starts with 100 and we want to get to the finish line with a total higher than our partner's."

3. Begin with a value of 100 on your calculator. As you cross a segment, perform the indicated operation on your calculator.

I grabbed a calculator and did a brief think aloud, "Hmm. I'm starting up here with 100 and I want to have a big number on my calculator by the time I get to the finish line. I guess I'll add seven-tenths. I wrote:

$$100 + 0.7 \text{ is } 100.7.$$

4. Move down or sideways (never up) through the maze from start to finish. You may not retrace any steps.

I slid my cube along the + 0.7 line to the next point on the board.

"Now it's your turn."

A student volunteer chose which path to move the class's game piece. Again we modeled starting with 100 and using the calculator to find the new total. I reminded the class that they also needed to write the corresponding equation on their recording sheet after each turn.

5. The goal is to choose a path that results in the largest value when you reach the finish.

After a few more turns I knew the students understood the mechanics of the game, so they were ready to play on their own.

"You're going to play with a partner," I explained. "Take turns moving along the game board. You can only move down or straight across, not back up toward the start. See who has the largest total when you reach the finish. Questions?"

"What if we get a repeating decimal?" Jason asked.

"Interesting," I replied. "If you end up with a repeating decimal, just round to the nearest hundredth." I knew students needed to be comfortable rounding to hundredths so I felt it was a reasonable and simple solution that didn't get us off track.

I asked for a volunteer to help hand out the calculators but the students informed me that they didn't need them. They already had laptops, tablets, and/or cellphones at their desks. I reminded them to only use the calculator function and not get distracted by other technological opportunities. I passed out game boards and centimeter cubes (to use as game pieces) and the students got to work.

The game was immediately engaging. Pairs of students jumped right in and began calculating and moving their pieces. A few minutes into the game a controversy arose. Some students were just following the exact same path as their partners. In this way they were assuring a tie at the end, but there wasn't any real mathematical thinking involved. Also, this technique annoyed the partner who was being shadowed. I interrupted the class to make a quick announcement.

6. Ask students to pay attention to their strategies and what they notice about operating on decimal numbers as they play the game.

"I'm noticing that some of you have developed an interesting strategy," I observed. "You're following in the exact same path as your partner. This strategy is clever because you know you won't lose. However, you're not getting an opportunity to think about the numbers and make decisions. So let's add a rule that you're not allowed to copy your partner. You have to make your own path. So if your partner is already on a number, you can't move on to the same number." To develop number sense, students must be thinking, planning, and developing strategies during the Decimal Maze game; not just copying their partner's moves.

There were a few groans and few sighs of relief. I took the opportunity to reinforce the idea that the point of the game was for students to think about the numbers and operations. I let the students know that we'd talk about their thinking and their strategies after they'd played a few games.

Play continued as Lynn and I circulated around the room. We asked students what they were noticing and why they thought it was happening. For example, when students observed that multiplying by a decimal resulted in a smaller product, we pushed students to explain what might be going on to cause this phenomenon. We also asked students to justify their choices and predict the effects of their moves prior to using the calculator. Reasoning abstractly and quantitatively requires attention to numbers, operations, context, and choices. The game provided all that.

We used questions such as, *What's your next move? What is the answer going to be close to? Why is that going to help you? Why aren't you choosing this move instead?*

These questions were vital in helping the students communicate their thinking and also clarifying the effects of different operations on decimal and whole numbers. One pair of students was struggling to make sense of the results of dividing by a decimal number. I decided to back up a bit and talk to them about whole numbers first. I wrote $6 \div 2 = 3$ on a piece of paper.

"So what does this mean?" I asked them

"It means you divide 6 into two groups," Shaleeni answered.

"Or it could mean how many 2s are there in 6," Mikael added.

"OK," I said, "Let's think about what division means with decimal numbers. If 6 divided by 2 means how many 2s are there in 6, what does 6 divided by 0.5 mean?"

"How many 0.5s are there in 6?," Mikael replied a bit hesitantly.

"Hmm. Let's see what that looks like," I responded as I drew six circles on the piece of paper.

Mathematical Practice

Students need to engage in the practice of making sense of problems and persevering in solving them (MP1).

Formative Assessment

Plan to ask a few questions ahead of time; this allows you to zero in on a few important concepts and thinking strategies.

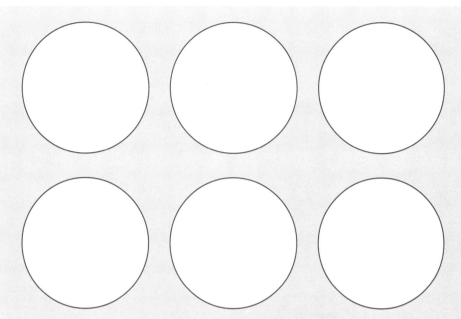

"With 6 ÷ 2, we can see how many 2s there are in 6."

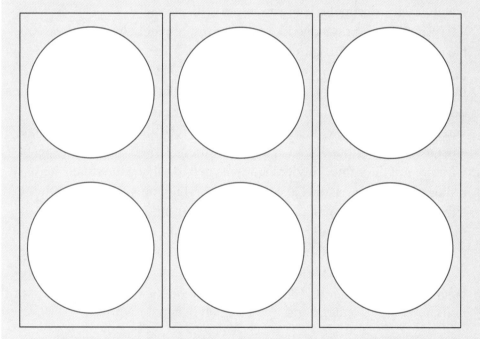

"I drew another set of 6 circles. So now we have 6 ÷ 0.5. How can we see how many $\frac{5}{10}$ s there are in 6?"

"Cut them into halves because $\frac{5}{10}$ is the same as $\frac{1}{2}$," Shaleeni offered.

I asked Shaleeni to demonstrate.

"So how many $\frac{5}{10}$ s are there?" I asked.

"Twelve!" Mikael burst out. "That's why the numbers get bigger when we divide the decimals. There are more little pieces."

"Ahhh," I acknowledged. "I wonder if that will help you when you have to decide on your moves for the *Decimal Maze* game."

I let the students play for about fifteen minutes and then interrupted to offer some challenges. This type of differentiation is vital for all students. Those who need more time and support can continue to play the version they started with. Students who are ready to move on can stay engaged by being pushed to apply what they learned to more complex variations of the game. I invited students to try to make the smallest total rather than largest or to work together to find the most efficient path for either the largest or smallest total possible. Another choice was to find the five moves that would yield the highest total.

After the students had played a while longer we got back together for a whole-group discussion. I began with a very open-ended question: *What did you notice when you were playing the game?* This led to observations about both strategy and operating on decimal numbers.

Teaching Tip

Extend the activity by offering additional challenges.

7. Go back to the original problems from the number talk and ask students to add to what they know.

I closed by going back to the original decimal problems I posed in our earlier number talk:

$$6 \times 0.9 =$$

$$15 \div 0.5 =$$

"Let's revisit these problems," I prompted. "There were some different ideas about these equations when we were talking about them earlier. Talk to your partner again and estimate the answer to each. See if you can use what you discovered in the *Decimal Maze* game."

"The answer will be lower because you're multiplying by less than 1."

"But it will be close to 6 because 0.9 is almost 1."

"I think $15 \div 0.5$ is a good move because the answer will get bigger," Shana predicted.

"How do you know?" I pushed.

"Because $\frac{5}{10}$ is the same as $\frac{1}{2}$. It's like asking how many $\frac{1}{2}$s are there in 15, and there are going to be, like, 30," she explained.

The students also reflected on their experiences in writing. As Tanya noted with amazement, *One of the things I learned was that when you divide a decimal the answer can get larger!*

Manuel wrote, *I learned more about decimals. What I have learned is that the decimal number can be big or small.*

Esme reflected, *It was so much fun to learn about estimating with decimals and the decimal maze was so much fun. I learned a lot from it. I love the way it stretched our minds to think about why multiplying a decimal by x will make x smaller.*

Lynn and I felt satisfied that the students were beginning to connect the meaning and effects of the operations on decimal numbers. Such understanding is vital to number sense. Rather than just using a procedure, students need to be able to have a sense of the reasonableness of their answers. Exploring how the four operations work with decimal numbers gives students a solid foundation.

Reflecting on the Lesson

Why did you allow the students to use calculators?

The *Decimal Maze* game is an exploration opportunity. Having calculators allows students to think about numbers and operations more broadly. It frees them from getting bogged down in procedures so they can use what they know about numbers, operations, and place value to make sense of their answers. The focus is on their mathematical thinking and reasoning rather than on crunching numbers and remembering rules. An important mathematical practice is using appropriate tools strategically. The use of a calculator for this investigation provides the appropriate freedom and flexibility to let students focus on the estimation and reasoning that are the focal points of this lesson.

How does this experience help students compute with decimal numbers?

Too often students are taught to compute with decimals by learning a bunch of rules and procedures. They count places, move decimal points, and perform other rote tasks without having much understanding of what they're doing or why they're doing it. Rather than rote procedural work, we want students to look for and make use of structure. The *Decimal Maze* game gives them time to explore operations and their effects on decimal numbers. These explorations and concomitant discoveries inform students' understanding of how our base ten number system works with decimal numbers. While students don't learn any procedures, they do learn to estimate, think about the meaning of operations, use mental computation, and make sense of decimal numbers as quantities.

References

Bresser, R., & C. Holtzman. 1999. *Developing Number Sense*. Sausalito, CA: Math Solutions.

———. 2006. *Minilessons for Math Practice, Grades 3–5*. Sausalito, CA: Math Solutions.

Burns, M. 2015. *About Teaching Mathematics*. 4th ed. Sausalito, CA: Math Solutions.

Chapin, S., C. O'Connor, & N. Anderson. 2013. *Talk Moves: A Teacher's Guide for Using Classroom Discussions in Math*. 3d ed. Sausalito, CA: Math Solutions.

Lempp, J. 2017. *Math Workshop: Five Steps to Implementing Guided Math, Learning Stations, Reflection, and More*. Sausalito, CA: Math Solutions.

National Council of Teachers of Mathematics. 2011. *Principles to Actions: Ensuring Mathematical Success for All*. Reston, VA: Author.

National Governors Association Center for Best Practices and Council of Chief State School Officers. 2010. *Common Core State Standards: Mathematics Standards*. Washington, DC: Author.

National Research Council. (2001). *Adding It Up: Helping Children Learn Mathematics*. Washington, DC: National Academy Press.

Trafton, Paul R. 1991. "Using Number Sense to Develop Mental Computation and Computational Estimation." Paper presented at a conference entitled Challenging Children to Think When They Compute, Queensland University of Technology, Brisbane Australia, August 9–11.

Reproducibles

R–1   Trail Mix Recipe   177

R–2   Get to 1,000 (Addition)   178

R–3   Get to 1,000 (Multiplication)   179

R–4   Decimal Maze Game Board   180

All reproducibles are available for download at www.mathsolutions.com/ mathworkshopessentialsnumbersense.

Trail Mix Recipe

You will need:

$\frac{1}{2}$ cup raisins

$\frac{3}{4}$ cup peanuts

$\frac{2}{3}$ cup granola

$\frac{1}{2}$ cup dried fruit

$\frac{1}{4}$ cup multicolored chocolate bits

Combine ingredients in a bowl. Mix well. Scoop into baggies for a snack on the go. Serves 6.

NAME _____

Get to 1,000 (Addition)

Ones, Tens, Hundreds

						Total

						Total

NAME

Get to 1,000 (Multiplication)

×1 ×10 ×100

									Total

									Total

NAME _____

Decimal Maze Game Board

Move down or sideways (never up) through the maze from *Start* to *Finish*. You may not retrace any steps.

Begin with a value of 100 on your calculator. As you cross a segment, perform the indicated operation on your calculator.

The goal is to choose a path that results in the largest value when you reach *Finish*.

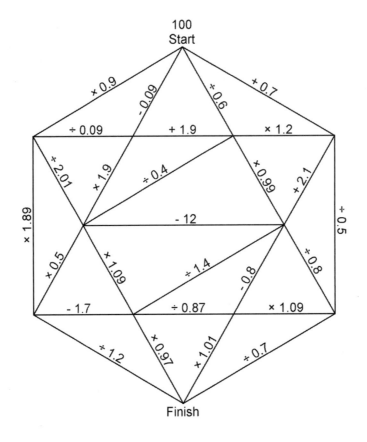

From *Math Workshop Essentials: Developing Number Sense Through Routines, Focus Lessons, and Learning Stations, Grades 3–6*. Copyright © 2018 Houghton Mifflin Harcourt. All rights reserved. Available for download at: mathsolutions.com/mathworkshopessentialsnumbersense.